A LIFE IN DEATH

Memoirs of a Cotswold Funeral Director

JAMES BAKER

A Life in Death
Memoirs of a Cotswold Funeral Director

First published in 2012 by
Mill Place Publishing

A catalogue record for this book is available from the British Library.

ISBN: 978-0-9573468-0-2

Cover design and typesetting by
Steven Levers Publication Design, Sheffield
www.stevenlevers.com

Contents

Dedication

This book is dedicated to everyone who appears in its pages. Some have been a small part of my story and others are a huge part of my life.

Especially to:
Mum & Dad, Sally, Granny, Nanny
and all my family
FMB & Family
CPW, MSA & JW
NT
DJM
RWJS
JN, CS and all the team, past & present
MS

And finally to MMT, who had always
threatened to write a book.
I thought one of us had better do it!

Author's Note

This is a memoir of my career to date. I have endeavoured to write it as sensitively as possible, balancing what I hope is an entertaining and interesting account with sufficient detail to enable the reader to gain a true insight into the working life of a funeral director. I have not sensationalised any of the subject matter and I certainly do not wish to cause any offence or distress.

The stories, incidents and events related here, which have occurred over a period of twenty five years, are entirely factual and very truthfully recorded. However, certain specific details have been omitted where necessary to protect the identities of the deceased persons and/or the bereaved families involved.

With the exception of the two funeral businesses under my own personal ownership - Lansdown Funeral Service of Stroud and Fred Stevens Funeral Directors of Nailsworth, together with the names of my two current full-time colleagues and my own family, all other personal names and company names have been changed.

Some readers – particularly local readers - may be able to identify certain high profile incidents related within this book and in those cases I have not repeated any information which has not already been in the public domain for a number of years.

CHAPTER ONE
A Strange Ambition

"I am not always good and noble. I am the hero of this story, but I have my off moments."
Love Among the Chickens (P.G.Wodehouse)

With gentle hills mottled in patchwork shades of green and timeless villages carved in honey-coloured local stone, the Cotswold countryside is a vision of natural beauty and rural calm. Yet, for one Gloucestershire student absent-mindedly contemplating the distant horizon as it shimmered in the warm spring sunshine, matters far removed from natural beauty and rural calm were occupying his thoughts:

I was sitting in classroom M2 at school, staring out the window and wandering what my first murder would be like. I would have to wait another three years to find out. Meanwhile, my English literature teacher was busy explaining the difficulties facing Shakespeare's Merchant Of Venice in trying to extract his pound of flesh. I would only have to wait a number of months for *that* kind of experience.

I'm sure I wasn't the first sixteen year old schoolboy to have pondered the practical realities of death in such detail as I did during that school term in the spring of 1987. But in my case these were not just the lurid imaginings of a teenage mind, because during the previous year I had already been baptised into the hidden world of the funeral profession. Even so, at that point in time I had still barely dipped a toe into the water of my chosen career and of course I had no idea that it would go on to give me so many fascinating and thought-provoking experiences; that I would find myself in strange and at times downright bizarre situations or that it would see me encountering deeply touching human moments as well as the most peculiar incidents of humour – albeit often unintentionally funny at the time!

Right from my very first day at the funeral home I felt an overwhelming sense of destiny. It has never been just a job to me, not even a career as such. It was, is, and always will be, my vocation.

Of course, I can't even begin to count the number of times people have asked me the question "So, what made you want to be a funeral director?" and they usually add the second question: "Does it run in the family?"

That second question always makes it sound like being a funeral director is akin to having a mental illness or being a serial killer, so my stock answer is to make light of it by saying that my grandfather was a School Attendance Officer, my uncle was a Church Of England priest, my father was an accountant and my sister works in teaching, so my family has established a precedent for doing the jobs no-one else wants to do!

However, that doesn't answer the main question – why did I want to be a funeral director? Ironically, given everything that's happened to me over these last twenty five years (and whatever may happen in the years still to come), I wish I could say my calling into the funeral profession actually started with an interesting incident, or some kind of "road to Damascus moment" perhaps. But no; the truth is, quite simply, that funeral work was just something I wanted to do.

To start with, I detested being at school and I wanted to leave at the earliest opportunity. The thought of staying on till I was eighteen and finishing sixth form, which nearly all my classmates would do, just wasn't something I even wanted to contemplate. With all the impatience of a typical teenager I felt as if the world was already passing me by and I was itching to get out into the world of work and adulthood, so I had set my heart on leaving school at sixteen.

Nevertheless, even I accepted that leaving school with no career prospects wasn't a viable option either. At the time I was actually interested in going into advertising, but even that line of work required a heavy stint in further education and so I was casting around for a different plan. Then, totally unexpectedly and in the space of an hour during an otherwise unremarkable weekday evening, a whole new vista of ambition opened out before me.

My destiny appeared in the form of a television documentary – a fly-on-the-wall programme following a Yorkshire teenager starting out as a trainee funeral director. My father, knowing of my preoccupation with all things morbid, suggested that I watch it. From the moment the documentary started I was hooked.

My most vivid memory of the programme, a powerful image that remained deeply imprinted on my mind, was a discreet long camera shot down a corridor, showing the apprentice helping to remove an unseen body from the mortuary fridge. Watching the scene left me with an almost unbearable feeling of curiosity.

The documentary fired my feverish imagination. The aura and mystery of funerals; working with dead bodies; I was totally sold on the whole concept. Here was a career that was made exciting by its peculiarity and something that I could really see myself doing. The fact that it was a profession where training was done "on the job", meaning it was viable to go into straight from school, simply added to the sense of "rightness" that I felt about it. From that moment my mind was made up - I wanted to be an undertaker. That was that and all there was about it. Little did I know that the chance for my newly hatched ambition to become a practical reality would present itself far sooner than even I could have dared to imagine.

It took me a while to admit to my parents that I had suddenly set my heart on a career in the funeral profession, as I was worried they might think it nothing more than a convenient excuse to leave school. But when I did pluck up the courage to tell them they took the news very well, all things considered. In fact it became my turn to be surprised when my father later came up with the suggestion that as he knew a local funeral director very well, he would ask him about the possibility of me having a look round his funeral home in the upcoming school holidays, with a view to letting me learn more about what the work involved. To my absolute delight, the funeral director in question not only agreed, but offered me a week's work experience – an incredibly generous offer, considering I might not even have lasted the first hour!

Not only did I survive the first hour, but I went on to spend my remaining school holidays at the funeral home before moving into full-time employment with the company for a further six years. At that time my new employers, a third generation family-owned firm, were one of the largest funeral directors in Gloucestershire, conducting over 600 funerals a year.

That was twenty five years ago and now I own my own funeral company. Naturally, over the years countless people have asked me how and why I became an undertaker and each time, after recounting my story, I finish by remarking that it's the only work I've ever done - a fact I often joke about by pointing out that I've never even had a paper round!

But before I take you on a journey through those very eventful years, let me also briefly set the scene. Like all funeral directors, I am a product of the locality which I serve and to understand my story you need to know where it has all taken place.

Tucked below the western escarpment of the Cotswold Hills, nestled in the meeting point of five valleys formed by the River Frome and its tributaries, sits the market town of Stroud. It sounds like the kind of rural idyll everyone thinks of when Gloucestershire is mentioned and in many respects it is. But as the telling of my story is true in every respect, so my description of its setting must be accurate as well.

Stroud is surrounded by the most picturesque, rolling countryside and whilst the town itself has many attractive buildings and quirky little corners, it is also a very gritty, workaday kind of place. It was a centre of local industry, originally a cloth-making town where the local mills were powered by the small rivers which formed the five valleys and from where the mills were supplied with wool from the sheep which grazed on the hills above. Only one cloth mill remains now, but they do make very high profile cloth products: the green baize for snooker tables and the cloth for tennis balls.

Stroud is also well known for having a significant artistic community. The town has been described as both "*The Covent Garden of the*

Cotswolds," and "*Notting Hill with wellies."* The second observation is particularly shrewd, because whilst Notting Hill may well be the last word in trendy bohemian living, you will find that in Stroud, as well as in London W11, the "boho" lifestyle is invariably underpinned by a level of affluence somewhat at odds with the values bohemianism would seem to represent.

Stroud – both a town and a district - is a relatively affluent place, but of course like everywhere else, it has its "haves" and "have nots" - a point illustrated by the shopping streets where independent niche-style shops share equal amounts of pavement space with the obligatory charity and pound shops, with both types of retail finding a loyal following. In fact, everything about Stroud speaks of vastly different worlds all existing cheek by jowl with each other.

But just a very short walk from the bustling normality of the town centre stood the portal into a very unique parallel world. It was behind the secluded but elegant façade of the district's leading funeral home that I would discover a whole new reality and from where, in the midst of death, I would see more of life than I could ever have imagined. It all began in the summer of 1986 when, as a fifteen year old lad lined up for a week of voluntary work experience, I took my first tentative steps into the unknown.

CHAPTER TWO
An Hour On The Day

"Undertaking? - why it's the dead-surest business in Christendom..."
Life on the Mississippi (Mark Twain)

Even before my week of work experience at the funeral home began, I would have my first encounter with the prejudice people naturally harbour towards anything physically connected with death. The incident caught me completely by surprise, but it was a valuable early lesson. It happened whilst I was spending a few days back in my native Wiltshire, where I was keeping a promise I'd made to my grandmother to do some decorating for her.

It was 23rd July 1986 - the day of the second royal wedding of the 1980's, when Prince Andrew married Sarah Ferguson and I recall seeing a few front gardens and houses decorated with union flags and bunting. Being a typical teenager I was of course totally disinterested in the royal nuptials and so as I walked into Melksham town centre to buy some more paint I was free to ponder matters far more dark and mysterious than royal weddings.

The following Monday would be my first day at the funeral home back in Stroud and the overriding thought occupying my mind, almost to the exclusion of all else, was the prospect of seeing a dead body for the first time. I had this inescapable sense that seeing a dead body would somehow be a life-changing moment; that the experience would in some way leave me changed for ever more. Yet, there was another part of me that couldn't wait for the moment; a part of me that, despite not yet even having any experience of that line of work, had already decided the funeral profession was the career for me.

But, as if the prospect of what the next week would hold in store for me wasn't daunting enough, I was also dragging behind me the reproach of my grandmother's disbelieving reaction to my intended career choice. Having not long before nursed my grandfather through terminal liver cancer, she couldn't begin to comprehend why I would now want to be involved with death and funeral work. My grandfather had lost his battle with cancer the previous year and of course that whole period of time was extremely hard for my grandmother, who had to witness her husband's descent from health and eventually from life itself.

I reflected on all of this as, with paint tins in hand, I headed back out of town, over the bridge crossing the River Avon and past the enormous Avon Rubber Company factory, where my late grandfather had worked as an electrical maintenance engineer. When Melksham folks referred to "The Avon" they usually meant the factory, not the river. My family had always suspected that the seeds of my grandfather's lethal cancer were sown during the legendary Great Fire at the Avon factory, in August 1966. The fire was the biggest in Wiltshire's history and had threatened to engulf the whole factory. It was an event which evoked a remarkable spirit of teamwork, with factory floor and office workers alike joining forces with the factory's own fire crew and the county fire brigade. Chains of employees, my grandfather among them, passed stocks of freshly made car tyres from hand to hand in the face of streams of molten rubber the inferno was creating. It wasn't hard to imagine the toxic legacy my grandfather might possibly have inherited from his part in this brave and successful mission to save the factory.

So I totally understood my grandmother's horror at my strange ambition and of course it made me feel very awkward about my intended career choice. To make matters worse I had no idea myself what I would actually be getting into and so it was very difficult to explain my reasons to her. But nevertheless, I could still sense the first, faint stirrings of destiny and my mind remained firmly made up.

Ironically, what neither my grandmother nor I knew at that point in time was that before very long she would begin to see a more positive side to my choice of career and in fact both of my grandmothers quickly became my main cheerleaders as I embarked on my unusual working life.

But in the meantime, those tricky days in Wiltshire were over again very quickly and my decorating tasks were finished. Leaving behind me the smell of fresh paint and the pleas of my still horrified grandmother, I headed back to Gloucestershire and prepared for my first day at the funeral directors, still convinced that I was on the verge of embracing an infinite mystery.

I had never noticed the presence of Thomas Broad & Son (Funeral Directors) Ltd., before, because despite their location on the very edge of the town centre, their premises were hidden away behind the trees that lined their driveway. In fact it was only the signs mounted on the banks at the bottom of the drive that told you there was even a funeral home there. My father had given me a lift down there and as the car climbed the steep entrance drive, Dartford House, a stately-looking, Grade 2 listed early nineteenth century pile, loomed up ahead, grand and imposing. Looking back, the qualities of that building perfectly reflected the purposes of its owners: invisible until you needed to visit it, but with a grand and reassuring presence when you did.

My father swung the car into the gravel car park and as I got out, a short figure in a dark pinstripe suit stepped out of the front door of the building. Standing on the pillar-sided porch, looking very relaxed with his hands in his pockets and his feet tilted so his weight balanced onto the sides of his shoes, the man in pinstripe greeted my father with a big smile and a very refined, clipped voice:

"Hallo John. You got time to come in for a coffee? No? Ok, right-ho, well leave the lad with us – we'll look after him!" The pin-striped figure turned to me. "Right my boy, you must be James. I'm Paul. Pleased to meet you. Come on in old son."

Saying a quick goodbye to my father, I meekly followed Paul inside. He didn't look in the slightest how I imagined an undertaker to be. In all the time I would later work for him he never would. Paul was the only person I knew who could wear the funeral director's usual conducting garb of either a tailcoat or a black Crombie overcoat, striped trousers and black tie and yet still not look like an undertaker. Somehow, I think he would probably have been quite pleased about that.

I entered a rather gloomy entrance hall, with stout, panelled doors on all sides and a side table decorated with a large, dried flower arrangement that would still be clinging to existence seven years later when, as a departing employee, I would eventually leave the building for the last time. Paul led me through a short and even gloomier passage immediately off to the right, which opened out into an office brightly illuminated by fluorescent strip lights and crammed with three desks. The room smelt strongly of coffee. Trying to take everything in, I was first introduced to Michael - Paul's right hand man – a tall, dark, broad shouldered chap in his thirties, and then to the secretary, Muriel, a large matronly lady with smiling eyes and half-moon glasses. (At later times in that first week, when I was left in the office if Paul and Michael were busy elsewhere, Muriel would be my ally – sharing with me her own knowledge and endless anecdotes about life in the firm).

There was a general hubbub of activity in the office and no sooner had we entered than the telephone started ringing. Muriel answered the incoming call whilst Michael had already picked up the phone on his desk to make another call on the second line. Glancing at Michael, Paul muttered a remark about it being a Monday morning and how much of a trial it was to get through to the crematorium, before he turned to face me. Leaning against Muriel's desk, he made a show of looking me up and down.

"Anyway, welcome aboard! This is the nerve centre of the whole place, but let's show you around the rest of the building and you can explain to me why you want to work in this beastly business."

Paul led the way back out into the passage and he flung open an adjacent door with a firm, brisk movement. Leading me into a small room with chairs set in front of a desk, he said,

"This is the interview room. Families come in here to go through the funeral arrangements, although equally often we will go out to meet people in their homes." Taking everything in, my gaze fell upon a framed, black and white photograph on the wall, a portrait of a venerable-looking old man in uniform. Hung next to his picture was an elaborate scroll-like document, referring to the Stroud Volunteer Fire Brigade. Paul must have noticed the direction of my gaze. "That's Thomas Broad himself. The old cove helped set up the local volunteer fire brigade many moons ago. Grim looking old bugger isn't he? Right, come on, let's show you the rest of the place."

With each step my feeling of anticipation grew.

We crossed the entrance hall again, towards one of the stout, panelled doors. As I passed the frosted glass partition that formed the rear wall of the hallway I thought I could see sunlight beaming through an open doorway some distance behind the glass. Although I was taking careful note of everything I saw, still everything was overshadowed by the thought of seeing a body. *We'd done the office and the interview room* I thought to myself, *so whatever comes next must surely be the nitty gritty? Would this be the moment?* I wondered anxiously. I followed Paul through the door into a large, slightly musty-smelling room.

For a split second, until I had a chance to take in my surroundings, I thought we had walked into a lounge. We had entered the room in its rear corner. Straight opposite me in the other corner was a wide sash window, almost floor to ceiling in height and finished with dull green curtains either side. The lower half of the window had frosted glass but I could see greenery and conifer trees through the top half and it seemed as though it looked out onto a garden. The room had a slightly tatty, dark green felt carpet and around three of the walls long wooden pews were positioned. In the middle of the floor was a pair of trestles with purple velvet covers fringed with gold braid. Behind the trestles, set in the middle of the front wall, was a small stone altar, with natural stone brickwork below, making the altar look for all the world like one of those fake stone fireplaces you can buy in d.i.y. stores. A brass crucifix and candle holders stood on the altar, although there was another stone cross set into the brickwork too. Above the altar hung a green tapestry and the whole feature was framed either side by tied back, yellow curtains. Paul, with his hands in his pockets once again, looked around the room as though inspecting it.

"This is the Chapel Of Rest".

It wasn't at all what I had imagined. I knew from the television documentary that a chapel of rest was only a viewing room, that funeral directors had mortuary fridges to actually keep bodies in; but I'd imagined that the chapel itself would be darker, lit only with candles perhaps. I had expected it to somehow be more dramatic and…funereal, I suppose. Ironically, with the benefit of hindsight I can think of few chapels of rest I've seen over the years that were any more funereal than the one at Thomas Broad & Son, but at the time I had nothing to compare it against, so my expectations were somewhat trumped by the seemingly mundane reality. Paul pointed at the stone altar.

"We salvaged that altar from the old town mortuary in Wotton-Under-Edge when they demolished the place, along with some old wooden head blocks…" *Head blocks? Old mortuaries?* My mind reverted back to the six million dollar question - *what will a body look like?* Paul carried on, "Now the coffin is brought in through this door", I turned to see another panelled door behind me in the corner, "and that way we can bypass the entrance hallway and move coffins in and out without any visitors seeing."

"Right", Paul gestured towards that second door, "Come on through".

Once again I steeled myself. *If this was the door that the coffins were brought in from, then it had to be…*

"The Bearers' Room" announced Paul, reaching round for the light switch.

As I stepped through the door, a windowless and rather tatty room was suddenly illuminated around me. I was beginning to see how the entrance hallway, the office and now the chapel of rest had all been converted from the original rooms in the rather stately building, although clearly they hadn't put in too much effort into the Bearers' Room judging by how old the paintwork looked. To my left, at the far end of the room, stood an old steel locker and down the length of both walls were rows of coat hooks, each hook laden with both a black raincoat and a thick black overcoat. *Death's locker room* I thought to myself, suddenly reminded of the changing rooms at school. To my right, next to another door, hung a clipboard which Paul pointed to.

"The bearers – the men who carry the coffins on funerals - are all part-timers, retired local men. When they're due in for a funeral we leave tickets on the clipboard detailing their upcoming jobs and they pick those tickets up when they come in here to get their coats. We'll put you on a funeral later in the week – one of the old perishers' coats should fit you."

"Do the bearers have a specific uniform?" I enquired, in a conscious attempt to demonstrate that despite my silence I was paying very close attention to every little thing.

"They come in with black trousers, white shirt and black tie and then we provide the coat," came the reply. "Mind you, we get arguments every year as to when it's time to swap over to Winter overcoats, or indeed back into mac's in the Spring. Silly old buggers! Right, onwards my boy."

He pulled the other door open and with another surge of anticipation I followed. With some relief I found myself in a flagstone floored corridor. At the far end, sunlight was beaming in through a pair of open doors and I noticed I was now stood on the other side of the glass partition which formed the rear wall of the entrance hallway. The partition and the door set into it obviously marked the divide between "front of house" and "behind the scenes" because out here in the corridor the décor was far more utilitarian, with the walls painted grey and all the doors finished in bright blue.

There was a far greater significance to this idea of "sides", that in time I would come to understand. That glass partition between the entrance hall and the rear corridor was a crossing point – both physical and metaphysical - between two worlds. It marked the divide between the hidden and the public, the separating point where the talk changed from "the body in the mortuary" to the "deceased in the chapel of rest". I would later spend the first years of my career mastering all that happened "out back," preparing bodies and coffins, before starting a different learning curve with "out front" duties, working with families on arrangements and conducting funerals.

Paul directed my attention to the floor of the corridor, along which slips of paper were neatly spaced against the skirting board, each slip with a surname written on in marker pen. Next to one of the pieces of paper was a cluster of wreaths, posies and a large cross of flowers, made out in white carnations with a red pleated ribbon surround and an arrangement of red roses in the middle.

"Florists deliver the flowers into here, next to the appropriate name tag, ready for us to take on the funeral. That cross should be the main floral tribute from the wife" Paul said, bending over and lifting his glasses up towards his forehead, squinting to look at small message card pinned to the cross. "Yes, that's it."

"Anyway, what makes you want to get involved with this whole mournful business?" asked Paul, still in that clipped voice, yet once again making a downbeat reference to his profession. I was puzzled; it seemed as if he really didn't want to be an undertaker at all. I would later learn that in fact he hadn't. He had fallen into it after marrying Thomas Broad II's daughter and the young couple had later been asked to take over the family business. Ironically though, it was on Paul's watch that Thomas Broad & Son evolved from one of many local builder/undertaker firms to become one of the largest and well known funeral companies in the county.

I explained about being inspired by the television documentary. Still anxious and hoping for some acknowledgement or reassurance about having to see a body for the first time, I went on to say that by doing some work experience in a funeral business I hoped to find out whether I really was cut out for the work. I mentioned about not having seen a dead body before and that obviously that would be the acid test - trying to make the remark sound off the cuff. Paul didn't seem to pick up the hint, saying instead,

"Our work is all about the living, James. We must hope to achieve some shred of comfort for the bereaved whilst at the same time arranging for the dignified disposal of the body. Our role on the day is to be a discreet presence, making sure everything goes smoothly. In this company we avoid all that ghastly theatrical business of marching around in top hats, wringing our hands like some character from a Charles Dickens book. Of course we need these premises and facilities, a fleet of cars and so forth, because there is much to be provided for a funeral and a lot of preparation involved, with the body, the coffin, etcetera. But fancy premises and cars are not an end in themselves. I say again, all of this is for the sake of the families and that one hour or so on the day as the funeral takes place. Our work is to provide a service – these premises and facilities exist simply to support that aim."

I was finding everything in the building so unutterably fascinating that I couldn't even begin to imagine that for Paul this was all just an everyday normality and a seemingly unwanted one at that. Another question popped up in my head.

"So what is the difference between an undertaker and a funeral director?"

"Nothing really. Undertaker is the old English term. Funeral Director is a dreadful Americanisation, but the term has gained traction in this country, so funeral directors we have become. Personally, I would much rather be referred to as an undertaker."

For all that Paul made that point, it would not take long for me to find that the word "undertaker" was never to be used in front of client families. It was to become one of many things about Thomas Broad & Son - and the people in it – that I found to be such a bundle of contradictions. It was only with time and through my own experience that I was able to unravel and understand all the things that, back then, had so puzzled my young mind when I trained and worked with the firm.

As Paul and I had been talking we had walked up to the other end of the corridor. We were stood by the open double doors, through which I could now see a very large yard, at the far end of which I noticed a hearse, parked halfway out of its garage. The sunlight was dazzling and warm summer air was wafting through the open doorway, but I was so over-awed and engrossed in this strange new environment that I'd almost forgotten it was still a normal Monday morning in the outside world.

The funeral director's premises I had seen on the television had seemed much more modern and plush than this and, considering the impressive frontage of Dartford House, I was somewhat disappointed with how plain and old fashioned it all seemed once inside. It would be some time until I was able to find out just how much better appointed it was when compared with many other funeral directors' premises back then. Meanwhile Paul gripped the handle of yet another large door, this one painted in the same bright blue as the yard doors next to it. He held the door open and beckoned me inside.

I had never – indeed never have since - encountered a smell quite like it. A completely other-worldly, disinfectanty smell, the like of which even now I can't really describe. Just very occasionally I will get a tantalizing whiff of it, just at some random moment, not even necessarily at work, but just something that recreates that smell and I am instantly transported back to that moment in 1986 - as if I have travelled back in time for a brief second.

The large room I found myself in was painted completely white, high-ceilinged, lit both by fluorescent lights and also by the daylight from a large window, again with the bottom half frosted. The floor was brown tiled – just like some of the old changing rooms at school and the walls were white-tiled half way up all round. There were two kitchen-style sink units with cupboards in the far corner, with a porcelain sluice in between them. If Dartford House had once been a grand residence, then this room felt like it could almost have been the kitchen – of the type where you would expect to bump into the cook, the butler or any one of the other household staff.

I say could *almost* have felt like the kitchen, because there were two things that destroyed the illusion. The first thing to catch my eye was a shiny, white porcelain slab. Rounded at one end, square at the other, the slab rested on two thick porcelain pedestals acting as the legs. The slab had a rimmed edge and across the surface were a series of shallow fluted channels that joined in the middle and led down to a plughole at one end, below which a pipe ran down into a grate in the tiled floor. The slab was a fascinating object, a perfect lesson in the cold application of form designed by function.

A subtle but funny human touch to this dread object caught my eye: in tiny letters at one end was the name "Twyfords." The sight of the tiny lettering instantly reminded me of the countless times I had contemplated that same word whilst stood in front of the urinals in the school toilets. (It was an unwritten rule at my all-boys school that you fixed your gaze firmly in front of you when in the communal toilets; to do otherwise risked accusations of being "queer" and consequently being thumped by a hastily convened kangaroo court). I suppose it stood to reason that the same factory that made porcelain urinals would also turn their hand to producing mortuary slabs, but even so I found the thought a strangely fascinating one.

The room echoed with a constant mechanical hum and as Paul closed the door behind us and I was able to see the rest of the room, I traced the source of the noise. In front of me, fitting perfectly into a large alcove was the mortuary refrigerator. The noise was coming from the motor unit sat on top of the fridge. Across the front of the fridge were three thick, insulated doors, each door with three little metal slots of the kind seen on filing cabinets. Nearly all nine slots had a little card in with a surname written in biro and in the corner of each card were strange mathematical sums like 6.3 x 20, or 5.8 x 18. By this time my heart was thumping hard. With no prior announcement or warning from Paul, the moment had come for me to embrace the infinite mystery of death.

"This is the mortuary." Announced Paul. "The fridge is designed to take nine bodies, but can accommodate twelve if needed. As you can see, we put these name cards onto the fridge doors so we know which bodies are where." He said, as he picked up a blank card from a pile on the adjacent window sill, I was rooted to the spot, my senses of sight and smell both overloaded, both battling to win the upper hand over each other.

"What are those equations written on the cards?"

"Equations? Ah, you mean the body sizes. I've never heard them referred to equations before. Equations! Ha! The body sizes enable the right size of coffin to be used. Take this one for example," he said, once again lifting his glasses and squinting at the small card in the middle slot on one of the doors. "Five eight twenty. Five eight means the length of the body: five feet, eight inches, whilst the twenty refers to the measurement in inches across the shoulders." He looked me up and down and declared "You'd be a five eight by eighteen, for example."

I was rather non-plussed by this revelation, more distracted now by the thought of the moment surely to come.

"Now then, old son, are you ready to see a body?"

"Um, yes, as ready as I'll ever be," I replied. "I won't get far if I can't handle that aspect of it," I said, trying to sound like it wasn't the one and only thing that had occupied my mind almost constantly for about the last month.

"This is true" agreed Paul, rather too solemnly for comfort, as he opened the far right hand door of the fridge.

As the door swung open, there seemed to be an even louder mechanical noise from inside the fridge, coming from the refrigeration fan inside. Inside, in a vertical tier, were three white fibreglass trays, each with a sheet-wrapped body on. Paul pulled out the lowest of the trays and pulled back the top half of the sheet.

It wasn't at all what I had imagined. The portly, elderly man's face wasn't white like I had expected, but simply a very pale, almost ghostly, shade of pink. The first thing that really struck me though was that he seemed to be smiling in a very absent way. His eyes were ever so slightly open and his teeth were all exposed, giving the appearance of a wide grin.

"This chap was a farmer from Painswick." Said Paul, naming a local village. "Now, are you coping ok, old bean?"

"Yes" I replied, spellbound but slightly repulsed by what I was looking at.

"Good lad" replied Paul, who carried on with an explanation that the body was still awaiting preparation and that when properly dressed and presented his eyes and mouth would be closed. I remarked at how it appeared he was smiling and Paul explained that the man's dentures had just slipped down, creating that appearance. The other thing that caught my attention was the white garment the man was wearing – it looked like it was made of a papery linen material, with a small ruffle surrounding the neck. It looked utterly bizarre and reminded me of a chorister's uniform.

Paul flicked the sheet back over the man's face and rolled the tray back into the fridge. I was fascinated and repulsed, but realizing I had survived the moment, I sought a distraction and remembered a sensible question to ask, prompted by terms I had heard elsewhere before.

"So what is the difference between laying out and embalming?"

"Laying out is simply the washing of the body, including plugging the orifices. The limbs are straightened, the hair combed and the body dressed. Embalming is a much more complex, chemical-based process, involving the replacement of the natural fluids in the body. That's carried out by injection of a preservative and disinfecting solution into the arterial system."

"So is laying out what they call last offices?"

"Good question my boy! It is indeed, although undertakers refer to "first offices" because it is what we do to the body first of all, whilst nurses refer to last offices as that is the last thing they do for the patient. That's a very well-informed question though. Where on earth did you pick up on all that?"

"I just read the terms somewhere." I replied slightly absently, once again lost in fascination at everything around me in the room. I still couldn't believe I was actually there at last - it all felt like I was having an out-of-body experience. I allowed myself a brief moment to think back to when I was decorating Granny's bathroom the week before and I reminded myself of what I was then thinking a body might have looked like. How wrong I'd been and how that previous week now felt like it had happened a lifetime ago!

My first time in a mortuary was all contained in just a few minutes, but I silently congratulated myself on surviving it, not only without feeling faint but also being sufficiently conscious to ask sensible questions at the same time. I followed Paul outside feeling pleased and extremely relieved, as if I had somehow justified his faith in allowing me to be invited into his funeral home in the first place.

We stepped out into the yard. That warm summer air felt fresher, sweeter and warmer than summer air had ever felt before. We turned left along the back wall of the building and back in through a set of sliding doors. We were standing just inside what was obviously the coffin workshop. This bit of the tour was much more like I was expecting: To my left were four rows of coffins, all stood on end and tilted back against the wall, each row five coffins deep, whilst to my right was a large wooden rack, three quarters the height of the workshop, containing slightly more elaborate coffins with panelled sides and different shades of wood finish. Further on past all the coffins stood a work bench, a pair of very aged trestles wrapped in hessian and in the corner what was clearly an engraving machine. The key cutting shop in town had a smaller, more modern-looking version which they engraved trophies and dog tags with.

The workshop had a lovely strong smell of woodshavings, even though there were none in sight, and after the mortuary I felt very safe in there. The coffins themselves were in a range of standard sizes it seemed, judging by the labels sticking out under the top end of each lid, with those same strange mathematical sums again: 6.0 x 18, 5.6 x 20. I remembered what Paul had said about noting the body sizes and rather satisfyingly it started to click into place in my mind.

"All the coffins are bought from our suppliers as finished shells, simply requiring us to add furnishings such as handles, nameplate and an interior lining. Three quarters of the coffins we use each year are a standard, simple design, like the veneered oaks behind you, whilst on the rack are the mid-range styles with panelled sides and raised lids." Paul said, indicating each group of coffins as he spoke.

He picked a coffin handle out of a box and held it up for my inspection. The handle flexed as he twisted it in his hands - a tiny shower of microscopic silver fragments falling away as he did so.

"Combustible plastic" Paul said as he let the handle return to its original shape. "We use them for both burials and cremations." Tossing the handle back in the box and brushing the glimmering silver fragments from his hands he continued "Electro-plated nickel finish, which as you can see is indistinguishable from real metal." He selected another random cardboard box, dug his fingers in and held up a little upside-down mushroom-shaped object, in the same nickel finish, with a screw projecting out of it.

"This is what we call a bell top. Once the lid is screwed down we screw these bell tops on top to cover up the screw heads. Like all the coffin furnishings it's combustible plastic, likewise so is the nameplate" he said, now holding up a rectangle of thin plastic finished once again in the almost mirror-like nickel coating.

Again, I concentrated hard, trying to take it all in; little did I know at that point, but in the following six and a half years that workshop would be my second home, wherein I would personally be furnishing nearly three and a half thousand coffins. Long before the last of those six years I would be able to fit and line a coffin in my sleep. I would be immersed in a world of Fleur handles in either nickel plate or electro brass, then, depending on whether the coffin was intended for burial or cremation and likewise whether for a Church Of England or Roman Catholic funeral, there would variously follow wreath holders, end ornaments, corner clips, 16" metal crucifixes, 8" filigree crucifixes and those dreadful "At rest" ornaments that always reminded me of Christmas cake decorations.

In time there would also follow those special occasions when a top of the range solid oak or solid mahogany coffin was requested and I would have to fit No. 305 style solid brass bar handles – always a fiddly procedure.

It's all changed nowadays. More and more now we're seeing coffins made from natural materials like woven willow or bamboo, or contemporary-style coffins of either wood or cardboard, often with printed, bespoke pictorial designs on them. Back in my first six years in the profession having an order for a solid oak coffin was a red letter day in that old workshop, but now it's more likely to be a colours-of-the-rainbow-dyed willow coffin or an unusual picture-finish coffin that raises my interest.

Looking back now, my fondest memory of the coffin workshop was when a local priest, who had a wonderfully sharp sense of humour, brought a young priest-in-training to look round, so that he could see what funeral directors do and learn more about the practical side of funerals. I was in the middle of furnishing a coffin for a Catholic funeral and as was the company tradition back then, the coffin lid was to carry a 16 inch metal crucifix with a plastic Jesus figure mounted on it. Before I had a chance to pin the crucifix in place, the older priest, who was already wearing a 3 inch crucifix pendant on a chain, snatched up the much larger crucifix from the coffin lid, held it against his chest, turned to the trainee priest and said,

"I'll hang onto to this one. It'll do for when they make me into a bishop!"

Leaving behind the workshop Paul took me across the yard to see the hearse I had spotted earlier – a Daimler DS420, of the kind so beloved by UK funeral directors for many years and of the kind most notably used for the funeral of Diana Princess of Wales. He showed me the firm's "removal vehicle" – a navy blue Peugeot estate car with matching navy blue curtains around all the rear windows. Opening up the tailgate, Paul pulled out a folding stretcher trolley of the style I had seen in so many films and he demonstrated its use. I was also shown the rest of the vehicle fleet, consisting in total of three hearses and three limousines. Within twelve months I would become intimately acquainted with every one of those vehicles, washing each of them umpteen times a week and in all winds and weathers.

So, with the guided tour completed, the scene set and my introduction to the world of the modern funeral director concluded, my work experience was to begin. Very quickly I would learn the reality of the point Paul was making to me on that first day: that from the funeral director's point of view the funeral is very much like the tip of the proverbial iceberg. There can be many hours work involved for a funeral that, very often, lasts for no more than an hour on the day itself.

CHAPTER THREE
Round To Rose Cottage

"He that toucheth the dead body of any man shall be unclean seven days."
Numbers, Chapter 19, verse 11 – King James Bible "Authorized Version".

The first task I was assigned to later in the morning of my very first day was to accompany Paul's colleague Michael to collect a body from Gloucestershire Royal Hospital. The nine mile drive took us up and over the line of hills that separate the Stroud valleys from the floodplains of the River Severn and down into the city of Gloucester. On the way Michael pointed out a tiny churchyard in the appropriately named village of Edge, which perches right on the very crest of the hill and has stunning views down to the river and across to the Forest Of Dean, sitting high on its tree-lined plateau on the western side of the county.

"If you've got to bury a relative, there aren't many spots nicer than that." Michael remarked. "Mind you, I did a burial of ashes there once and it absolutely pissed down with rain. I could see the storm coming up the Severn Vale and I was silently wishing the bloody vicar would get on with, but he didn't and we got soaked!" he chuckled.

I wondered what was involved with burying ashes but Michael was already chattering about something else and I didn't get the chance to ask.

On an altogether less funereal note, before that first week was over, if Michael mentioned "shagging" once, he must've mentioned it a hundred times. I hadn't heard anyone use that word since primary school back in my native Wiltshire, when a boy called Matthew, a scruffy waif perpetually dressed in a tatty jumper with sleeves too long for his arms, would delight in repeating jokes that his much older brother Shane had told him. I didn't really understand the jokes but often there were references to animals as well as s***g, creating an association in my mind such that even now I still hate hearing the word. But my feelings aside, Michael clearly had no such qualms and often in times to come, when riding out with him to removals or other errands he would gleefully recall tales of his much younger days when, as a local jack-the-lad, he'd go out and go...well, you get the point.

He went on to ask if I was s***g, at which point I shifted uncomfortably in my seat and admitted that I'd been at an all-male school and hadn't had the chance to meet many girls (and trying to avoid admitting that, truth be told, I barely knew one end of a female from the other). Looking back, I can see why Michael would have been a hit with the women: he was tall, well-built, dark haired, with boyish looks and an ability to charm people in general, and low-hanging female fruit in particular, with equal ease.

The journey itself to Gloucester was one I was familiar with. But knowing the purpose of our trip and riding in an undertaker's removal vehicle for the first time, with the stretcher trolley rattling just behind my seat, I felt as if I had somehow entered a parallel universe. By its very nature the funeral profession is very discreet and low key and like so many other things in life, it's functioning day to day just like everything else; it's just that we never notice it.

Of course, I had still only very briefly seen that first body and I was already starting to feel anxious again about seeing another. What would this one be like? Was successfully surviving the first encounter just a fluke and would I buckle this time around? Despite Michael chattering away as we had travelled into Gloucester I had been mulling over thoughts of nervous anticipation and now the moment was once again nearly upon me. Michael turned into a rear access road to the hospital and we negotiated speed humps as anonymous-looking buildings with signs like "Estates Department" and "Cotswold Dialysis Centre" passed by the car window.

Gloucester has two landmarks visible across most of the city: the magnificent tower of Gloucester Cathedral and the utterly un-magnificent main tower block of Gloucestershire Royal Hospital; a huge, unforgiving monument to 1970's dystopian concrete architecture that now loomed heavily over us as we drove through the hospital grounds. I was nervous enough as it was and the tower block made me feel ill just looking up at it.

A couple of hospital porters in an electric milk-float type vehicle, the back laden with laundry sacks, pulled out onto the access road a little further ahead of us. We were soon stuck behind them, but Michael said "No problem, we're in here anyway." as he suddenly swung the car into a turning hidden behind some conifer trees.

The turning curved tightly into an immediate dead end behind the trees, after which Michael reversed the car back a few metres and turned off the ignition before the car had even stopped moving. He got out of the car, but the way he had pulled in and parked made it seem like he was just stopping there for a moment. Unsure of what to do I decided to get out too. I turned to see that the rear end of the car was parked under a covered bay, with double doors immediately behind. To the side of the doors was a large blue sign with friendly-looking white lettering, very different to the normal style of hospital signs:

Mortuary Opening Hours
11.00am – 4.45pm
Closed 1.00pm – 2.00pm
Ambulance staff or funeral directors wishing to gain admission outside these times please contact the hospital switchboard

For the second time that day I had been caught unawares by the sudden presence of a mortuary. Michael pressed the button on an intercom set into the wall. A tinny-sounding voice emitted from the speaker in response to which Michael leant closer to the intercom and responded:

"Hiya. It's Michael from Thomas Broad's."

A few moments later there was the sound of a door latch being turned and one of the doors swung open, to reveal a man in his thirties, dressed in medical scrubs and with a blonde, mullet-style haircut. He stood in the doorway, a wide grin across his face.

"Hi Mike. You bringing in or taking away?"

"Just taking." Michael replied, pulling the stretcher from the back of the car and letting its trolley legs fold down into position. He pulled the stretcher through the open door, with me following behind.

Michael explained my presence and the mortuary attendant (or "technician," to use the proper title, as I would later learn) introduced himself as Ian. After politely responding to his greeting as I nervously stepped through the front door, I took in my surroundings and found myself stood in a large vestibule area. That unmistakeable smell of hospital hung in the air and at first glance it all looked like we could have been standing in virtually any department of the hospital. However, through some open double doors to my left I could see a room with a large bank of fridges taking up one wall. At the far end were more double doors with narrow frosted glass windows and no entry signs. The sight of the fridges was much more like I was expecting a mortuary to be.

"Which side are we on?" Ian asked, as he took an A5-sized paper slip from Michael's hand.

He glanced at the form he'd been handed.

"Gibbons. Right ho. She's hospital side."

"Ooh ooh ooh the funky Gibbon" this Ian character muttered to himself as he ambled around a corner at the far end of the vestibule area. He had the cock-sure air of someone who really fancied himself.

I followed Michael round the corner and was confronted with another bank of fridges identical to the first. Likewise, at the far end of the room was another set of double doors with no entry signs. (I was later to learn that this was a public mortuary, in the sense that as well as hospital deaths, it also received bodies brought in from all around the county whose deaths had been reported to the Coroner. Thus one bank of fridges - each bank taking 30 bodies - was reserved for deaths in the hospital, whilst the other bank was used to receive coroner's cases brought in by funeral directors or ambulance crews). Opposite this second bank of fridges stood what looked like a huge red weighing scales, with a small sticker on the front saying something about deducting a certain weight for the trolley. I noticed the section of floor I was standing on was wobbling slightly beneath my feet and I realized the scales were built into the floor.

Ian glanced at a large, bound register book on a shelf and then selected a fridge door. As he swung the heavy insulated door open the room was filled with the mechanical noise of the refrigerator fans, just like when Paul had first showed me his own mortuary. There were four body trays in a vertical row behind the door this time. Ian pulled one of the galvanised metal trays out a short way and then manoeuvred a large scissor-lift trolley with rollers across the top of it into a position ready to receive the tray. Meanwhile Michael had been taking the cover off the stretcher and undoing the two straps. Ian yanked on the end of the metal tray and pulled it out of the fridge and along the length of the lifter trolley. The tray made a sound like rolling thunder as it passed over the metal rollers inside the fridge and then a clank as it reached the stop bar at the end of the scissor lift trolley. It all seemed so noisy and industrial and I noticed the equipment was a lot older than it was in Thomas Broad & Son's mortuary.

As Michael positioned the stretcher alongside the tray, Ian tugged at the white sheet wrapped round the body. I braced myself for the sight of my second dead body. There lay the thin, almost emaciated body of an elderly woman, wearing that same bizarre white shroud with the ruffled neck as the man I had seen earlier. Her head, with nothing to support it, lay back at an angle, her hair long and straight, hung back from her face. Her eyes and toothless mouth were wide open. She looked as if she was locked in a silent scream. Her whole appearance was considerably more jarring than the elderly man I had seen a couple of hours previously. But it was her eyes that I noticed most: misty-looking and crumpled, in the way a ping pong ball is when you dig a thumb into it. Ian must have noticed me looking.

"You've seen a body before?" he asked.

"Um, yes, but only one." I said, "But I was just looking at the eyes. They seem sort of crumpled-looking."

"Dehydration. The eyeball is normally fluid-filled, but after death that fluid dehydrates and you often end up with that crumpled effect."

"Riiight" I said slowly, processing this new knowledge and looking back at those dead eyes again. (In the days to come I would learn that the problem was solved by the use of "eye caps" – basically thin, plastic hemispheres like giant contact lenses, that are inserted under the closed eyelids to give the eyes a more natural, rounded appearance).

Ian pulled one sleeve of the shroud to expose an identity wrist band and looking at Michael, he said,

"Doris Gibbons. No jewellery. There's a tub with her dentures in though."

Michael looked at the wristband.

"Ok."

Ian tilted the tray from one side and Michael pulled the sheet, sliding the lady's body across and onto the stretcher. He flung the sheet back over the lady's body and clicked the two straps to hold the body in place. There was that sound of rolling thunder again and a ringing clang as Ian pushed the empty tray back into the fridge and let it roll back into its berth. He swung the heavy fridge door so hard that I braced myself for a loud bang when it shut, but instead there was just a solid, muffled thud. All the while Ian and Michael carried on chatting cheerfully.

"Where's Gerry then?" Michael asked.

Ian cocked a thumb towards the "No entry" doors.

"Still finishing off in the PM room."

PM? That must be how they refer to post-mortems, I thought. The very utterance of the term "PM room" sent a nervous thrill straight through me, as I thought for one moment that Ian might open the door and that Michael might go in there and expect me to follow. I was still feeling completely wrong-footed by everything and to my great relief the doors stayed unopened. Instead Ian was holding his finger next to an entry in the register book, waiting as Michael signed and then left a pound coin on the book, which Ian scooped up and dropped into a jar in a drawer beneath.

Michael swung the trolley round, picked up the elasticated cover and beckoned me to take one end and fit it back over the body. Ian held open one of the front doors again and Michael wheeled the stretcher out and folded it back into the car. There was a white Bedford van parked in front of us, tucked up into the turning area. I could see the wording "Private Ambulance" in red letters down the side of the van.

"And still they come," said Ian.

"Yeah. No rest for the wicked, eh Ian?!" answered Michael.

"Plenty of them here!" Ian chuckled. "See you, Mike. Goodbye James."

It seemed strange to hear Ian say my name.

"Goodbye" I replied, my thoughts still behind in the mortuary even as I got back in the car. There was an air of quiet menace about the place, inside and outside, grimly enhanced by a soundtrack from the constant, all-pervasive, mechanical humming of the plant machinery that powered the enormous fridges. Looking back, that feeling of menace was simply a natural reaction to not being used to bodies or mortuary environments. Having seen two bodies I had realised how much it would take me just to get used to that, let alone the distant challenge of seeing a post-mortem or a badly injured body. I thought back to Ian again, wondering what he actually did, what things he actually saw, working in a mortuary all day long. I had already learnt that mortuaries and bodies were just one part of a funeral director's work, so I wondered how Ian coped with working solely in a mortuary environment all the time. I would find out in time, because within a couple of years I too would no longer be able to remember what it was like not to see dead bodies on a daily basis.

The journey back was also a little odd, as all the while I was conscious of the body just inches behind me, the stretcher rattling and shifting slightly as we drove along. Michael and I chatted about various things, with me asking questions about funeral work and the mortuary we had just left, but Michael more interested in knowing a bit more about me and what my school was like. He had gone to a local comprehensive school barely a mile away from the public school at which I was still a pupil.

Looking back it all seems rather funny: there I was, full of questions after my first few hours in that grimly fascinating parallel universe, whilst Michael was equally curious about what life was like in a "posh school"; the two of us trading questions, probing with almost childlike curiosity about what it was like in each other's world. The conversation switched subject again when, as we were travelling along the road in which Thomas Broad & Son's funeral home was located, Michael pointed out a woman in her late fifties walking along the pavement with shopping bags in hand.

"I've done a funeral for her. Her husband."

I wish, over the following seven years, I'd been given a pound for every time he would point someone out and say "I've done a funeral for him/her"

The following day I accompanied Paul to a further two hospitals at Cirencester and Stroud - both small town general hospitals. This time around though, both the mortuaries were small, old-fashioned outbuildings with no permanent staff. Instead hospital porters who would come out to unlock, assist us with loading the body and then ensure that the mortuary register was signed. But it was on that second day of work experience, at Stroud General Hospital, that I was to actually touch a body for the first time.

We'd parked outside the mortuary - an anonymous-looking Cotswold stone building at the rear edge of the hospital campus. The first task was then a short walk back down to the rear entrance of the main building and through to the reception desk to ask if there was a porter around. The receptionist put out a call whilst Paul & I threaded our way through the corridors to the porters' room, where the mortuary register was kept. Less than a minute after we walked into the room, a porter appeared and pulled from a shelf a large, bound register book identical to the one I had seen at Gloucester. Paul signed it and handed over another of the little A5 slips of paper – a "body release form," he explained to me. The porter selected the key to "Rose Cottage" as I was to discover the mortuary was discreetly referred to in all small hospitals and we walked back up there.

Inside, it was an open, bare space, the walls of the same un-rendered Cotswold stone as the outside, with a crucifix hung on the far wall. There were no fridges, just four curtained cubicles, each with a low trolley on which the body, wrapped in a sheet, would be laid before being draped with another sheet. There was no air conditioning or temperature control and as I was to learn in times to come, that meant autumn and winter were good times to visit, but summer could be a different matter. There was a rather unpleasant, musty smell in the place as we entered. In all the innumerable visits I was destined to make back to that mortuary over the next seven years that smell was never to change, it just got mustier in the summer times! In the event, there was only one body there, but even so Paul and the porter still went through the process of checking the identity tags. The significance of that was not lost on me. Paul then folded the stretcher down to floor level and said,

"Right lad, take one of the arms."

That caught me out a bit. Without having a chance to think about it, I braced myself and took hold of the left forearm, aiming for the sleeve of the shroud to avoid touching the skin.

"Use both your hands, old son, take him by the hand as well."

Already holding the dead man's arm aloft with my right hand, my left hand gingerly grasped the dead hand now hovering in front of me. I cringed inwardly for a second, struck first by the sensation of holding a hand which wasn't going to hold mine in return. It was a spooky moment – wondering whether the hand might suddenly grab mine! It wasn't warm, or cold, just... room-temperature. It was slightly clammy, but what I noticed most was the way in which the hand didn't move. The forearm moved in my hand as if hinged at the elbow, but there was no other flexibility. This was my first experience of the effects of rigor mortis.

I have been asked about rigor mortis innumerable times, so let me explain it quickly now. It is caused by a chemical change in the muscles after death, causing the limbs to become stiff and difficult to manipulate. Broadly speaking, onset occurs in the first twelve hours after death and the effect begins to dissipate in the following twelve hours. In practice this means that when rigor mortis is at its fullest extent, if you can even manipulate a limb at all it will instantly assume the same position it was in. After twenty four hours, that same limb will still be stiff to move, but with gentle and gradual manipulation the limb will become flexible and will subsequently remain so.

With the body now on the stretcher I felt quite elated. I had seen bodies and now I had touched one. More importantly I had managed these things without feeling sick or queasy. Although, in my initial enthusiasm after seeing the television programme I had already imagined myself as an undertaker, I knew now that I really might actually be capable of doing the job.

So far I had been given a good insight into the "behind the scenes" aspects of the undertaker's work, becoming so absorbed in it all that I had almost forgotten about the actual funerals themselves. The latter half of the week however, brought my first experience of being "front of house," when I actually went out on a funeral.

It was a damp, grey Thursday afternoon. The funeral itself was to consist of a service in a church within the town, followed by burial (or interment, as I later noticed it was referred to in proper terms) in the town cemetery, just along the road from the church. I was just there to observe, but Paul told me to station myself by the church door ready to take the bunches of flowers some of the family mourners were clutching. The responsibility of even this simple act weighed heavily on me. Once the procession had entered the church and everyone had been settled, the service began. Paul and the bearers came out of church, gathered up the flowers I had collected by the church door and put them in the hearse. The bearers remained outside with the vehicle whilst Paul took me back into church and we stood at the back listening to the service.

The one abiding memory I have is of hearing the vicar talking of the life of the deceased and, with the mention of the person's name, realising that it was the body that Paul and I had removed from the hospital two days before – the first body that I had touched. It suddenly hit me that the stiff, lukewarm, clammy body that I had been so hesitant to touch was in fact the remains of a once living human being: a husband, father, grandfather, a keen gardener with a well tended allotment and an avid football fan, according to the vicar.

Listening to the vicar's words was the moment when I finally made the connection between the "bodies" I had seen and the fact that they were someone's loved one. It was a significant moment – another turning point. The service drew to a close and the short trip to the cemetery was made.

I stayed by the hearse whilst the burial was carried out and with the family lingering by the grave looking at the flowers, the bearers and I climbed back in the hearse, after which one of them pulled out a bag of sweets and offered them round. The bearers started chattering amongst themselves and comparing tickets to see who was doing which funerals the following day. Although they were all retired, this was their daily reality. I, on the other hand, would soon be back at school and I remember feeling incredibly jealous of them.

My week of work experience drew to a close, but to my delight I was offered the chance to come back to the funeral home during the next school holiday. This pattern was to develop into a regular fixture for the remaining school holidays until I left school the following summer. All of my school friends were of course only too glad of the break, but I spent all my time helping out at the funeral home and it was during those school holidays that there were to be two more important experiences for me.

The first concerned a body that Paul and Michael had removed from a private residence. At that stage I was only allowed to go out with them on hospital removals. Removals from private residences and nursing homes, because of the unpredictability of what may be waiting, were off limits to me at that point. As they wheeled the stretcher across the yard from the removal vehicle, I followed them into the mortuary. They removed the cover from the trolley and unzipped the flexible stretcher beneath (the flexible stretcher is basically a heavy duty body bag with six handles and with wooden slats fitted into its underside to support the weight of the body. It is used to remove and carry the boy from confined spaces to an area where it can then be placed onto the main stretcher trolley).

The body was that of a man in his sixties, naked, still warm and completely yellow. A bilious, cancerous yellow. I started to help them lift the body onto a fridge tray, but already my mind was telling me that this was too much. Just seeing all that yellow, jaundiced skin was proving a little hard to handle. They could see I was struggling and told me to go outside if I wanted to. I gladly took them up on the offer. I stood outside the mortuary door. For a brief moment I was glad I had admitted defeat, but as the day wore on I became really angry for letting myself down. Paul and Michael were absolutely fine about it, very understanding, but I felt as if I had somehow failed and I vowed to myself that I wouldn't let that happen again.

The second incident was my first encounter with what, for now, I shall refer to as a "tragic" death. I use that word carefully here because, as I will explain later, I would go on to learn that the assumption some deaths are, by their very nature, more tragic than others, dangerously undermines a far more fundamental truth.

Michael and I were off on a long distance hospital removal. A gentleman had drowned whilst on holiday at a seaside resort with his family. The reason the incident sticks in my mind was that the deceased was the same age as my father and the knowledge that some of his children, all roughly my age, had tried desperately to help him in the water until the emergency services arrived made a deep impression on me.

The journey itself took about two and a half hours, with a rest break on the way in a Little Chef. Michael stressed to me that normally the firm would only send one member of staff and so any rest break had to be made on the way there, not on the way back with a body on board. Stopping for petrol was one thing, he explained, but leaving a vehicle and body unattended in say, the car park of a motorway services, just wasn't an option. I enjoyed riding out with Michael – in his spare time he competed in autograss racing and whilst his normal driving wasn't dangerous, he certainly didn't hang around either.

We arrived at the destination hospital – a large city general, and began looking for the mortuary. In later years when I was doing such removals myself, in addition to route planning in general I would always obtain directions to the mortuary in advance. Mortuaries are intended to be inconspicuous and are often hidden in all kinds of anonymous nooks and crannies within hospital grounds. Only once have I ever driven through a hospital front entrance and seen the mortuary listed on the main sign along with every other department. So many times in the past I had cursed mortuaries that were difficult to find, even with directions, and yet when I did finally see that one signposted right from the front gate, not only was it somewhat surreal but I actually thought it was rather tactless…!

On this early trip with Michael however, luck was on our side. Becoming ever more conscious of anything funeral-related, including learning to spot undertakers' removal vehicles, I suddenly noticed a likely-looking black Ford Granada estate pulling out of an entrance down an access road.

"Is that an undertaker there?" I said, pointing at the vehicle pulling away.

"Yeah, I think it is. Well spotted James. You're learning! Let's turn down there then and have a…yeah, bingo. There's a sign - parking for funeral vehicles and ambulances only. That's the place. Right, let's see what they've got for us then."

Once again, there was that vaguely menacing mechanical hum echoing around the place and as Michael pressed the bell I waited with anticipation to see what that particular mortuary would be like.

Apart from a cursory glance as Michael checked the identity tags I didn't really get to see the body until we were back at the funeral home again. I felt a little uneasy on the journey back, wondering if the body just behind my car seat was somehow more infectious because it was a death by drowning. With the benefit of experience I know now that the risk of infection is the least thing to worry about with drowning cases. The only thing for us to be worried about is not so much that the person has drowned, but how long their body has actually been in the water. That can make a difference - a very big difference, as I will explain later…

When we returned to Thomas Broad & Son, transferred the body onto a fridge tray and unwrapped the sheet, I was instantly struck by how human the body looked. Up until then all I had seen were the bodies of elderly people. I was gradually getting used to often very thin, fragile bodies, with wizened features and fly-away, soft white hair. But here was someone, who, but for the fact that he was dead, looked almost like he could have got up off the tray and walked around. A discernable hair style, a beard and a face as full as it was peaceful, all atop an otherwise healthy-looking body, with muscular arms and firm, thick hands.

I've never been one for the "that could have been me or mine" kind of sentiment, but even so, here for the first time was a body I could relate to. I imagined his children, roughly my age, losing their father and I tried to imagine being in their position. It was a very unpleasant feeling and I quickly put the thought from my mind; but the moment left a deep impression on me. Having got lost in learning the many different practicalities of the work, here suddenly was another sharp reminder to me of the human reality at the root of all that funeral directors did.

Not very many months after, I would be standing there in the mortuary again, handling the body of a young father who had lost his battle with terminal illness. Accompanying his body were some pictures painted by his children, together with a little figure one of them had made at school, using a toilet roll, pipe cleaners and coloured tissue paper. Another very poignant moment that once again served to remind me of what lay at the heart of the work I so wanted to be a part of.

As I have said, I was invited to return in subsequent school holidays and it was during one of those periods that Paul asked me into his office and told me that there could well be an opening within the company for me as a trainee. At that stage it was just him and Michael doing most of the work, with some part time help from one or two of the more capable bearers. In due course Paul told me a plan to fund my recruitment to the company had been agreed with their accountant and if I was still interested I could start work with Thomas Broad & Son as soon as I left school.

Paul explained that my employment would be funded via the government's Youth Training Scheme (YTS). The YTS training courses lasted for one year, later amended to be extendable to a second year. This extension to a second year would have useful ramifications for me by the time I got to that stage. My parents' house was less than a mile away from the funeral home, halfway up a steep hill and Paul, who lived further up on the top of the hill, would always give me a lift home each day. But after that job offer - which of course I accepted on the spot - I was in such a cloud of elation I think I could have floated back up the hill to my parents' place. My destiny was sealed. I was on my way.

Not long after the job offer, when I was back at school again, I was required to sit a careers aptitude test. In addition to the tests there was a question asking pupils to express an interest in any particular career. Naturally I wrote "Funeral Director & Embalmer." A few weeks later when all the tests had been sent back from marking, I was summoned to the careers master's office for my appraisal. He was, he said, most disgruntled to receive a letter back from the school careers testing organisation along with my test results, saying that although my results demonstrated a good degree of ability I clearly wasn't helping my cause by not taking the test seriously, by saying I wanted to be a funeral director / embalmer. As any 16 year old would, I went on to take great delight in proving myself innocent of the charge, before stating proudly that I already had a job offer as a trainee funeral director. The careers master was very magnanimous about it and after that we darkened each other's doorsteps no more!

CHAPTER FOUR

"Ah, so you're the one who wants to be an undertaker"

"Everything dies, and on this spring morning, if I lay my ear to the ground, I seem to hear from every point of the compass the heavy step of men who carry a corpse to its burial."

Madame Gasparin

I started work full-time on Monday 6th July 1987, having already done my two day induction as a Youth Training Scheme trainee. I would have to spend one day a week at an approved training centre – in my case the YMCA in Cheltenham, where I was assigned to the "Health & Care" course - designed primarily for those wanting to work in care homes or medical settings, but the only course even remotely applicable to a trainee in the funeral profession.

There was a television advertising campaign running at the time, featuring trainees from around the country, ending with the line "The best advertisement for YTS is the people who've done it". They never asked me to appear in one of those adverts, strangely enough. However, I later discovered that I had achieved a tiny amount of fame within the Cheltenham YMCA, when during one training day the course tutor told me I was wanted down in the administration office for some minor paperwork check. I reported to the office and waited for the administrator to retrieve my file. "James Baker, James Baker" she murmured, searching through the filing cabinet. She eventually located the file, glanced at the cover and then, fixing me with a curious stare, said,

"Ah, so *you're* the one who wants to be an undertaker."

I stood there and smiled innocently.

Anyway, let me go back to my first day at work, as there is another character to be introduced: Rick – 3 years my senior, who was also taken on by the company just a short while before me, as another funeral assistant. Rick and I would go on to become – and indeed remain – best friends. Like me, Rick had long harboured a desire to work in the funeral profession, although in his case after a short spell of part-time employment with Thomas Broad & Son he had gone on to secure a position as a funeral assistant with the Gloucester branch of a well-known national company.

However, he had been invited to re-join Thomas Broad & Son on a full-time basis a few months before I too started with the firm as a proper employee. Rick was a very quiet, reserved character, although as I got know him better I found much truth in the old saying about still waters running deep.

Paul would in later times describe Rick as "An enigma, even to his own mother." I recall offering the thought that he was just "A very self- contained sort of bloke," whilst Michael's considered opinion was simply that Rick was "A weirdo who just needed someone to give him a good shagging," before going on to suggest a couple of the female staff in various local florists' shops.

When I was first introduced to Rick we shook hands across a body in the firm's mortuary. I can remember the dead body in question perfectly clearly – a young man in his thirties, although I confess I can't remember what actually led to his premature end. I think it was an accident of some kind.

Rick had just finished preparing the body and was placing the lid back on the coffin, ready for it to be loaded into the removal vehicle. Paul then asked me to accompany him in transferring the coffin down to the company's branch office in the district town of Dursley, ten miles south of Stroud.

I had been down to the Dursley office once or twice during my periods of school holiday work experience, so I was familiar with the

Carry On film-style drill for getting coffins and stretchers in and out of the premises. The branch office was situated on a corner site half way up a busy street. Although there was a single parking space behind the office for the branch secretary to use, there was no rear access. Instead we always had to park at the bottom of the steep lane that ran up the side of the office, look around for any pedestrians and, when the coast was clear, pull the coffin or stretcher out of the car and carry it in through the front door as quickly as possible. During my time with the company the Dursley branch would be refurbished and a much needed rear access added, albeit much to the annoyance of the neighbouring bakery whose delivery access was right next to where we were then bringing coffins and stretchers in and out.

Having started work officially now, a daily routine quickly developed for Rick and I. Every morning we'd start by removing the coffins for the day's funerals from the fridge and sealing down the lids. This involved using a large ratchet screwdriver to fit four long wood screws per coffin lid. The only thing more innumerable than the number of coffin lids I must have screwed down over the years was the number of unrepeatable swear words I shouted every time the huge screwdriver slipped – either scratching the coffin lid or, worse still, impaling my fingers. Once the day's coffins were sealed, the car washing would start; and with a total fleet of eight large vehicles it seemed a never-ending task. I detested it.

With the car washing completed, I would then be kept busy in the coffin workshop fitting and lining coffins whilst Rick would go off and do hospital removals. This left me to establish a new workshop ritual of downing tools at eleven o'clock every morning and listening to "Our Tune" on the Simon Bates Show on Radio One. I would get constant ribbing from Rick about listening to "Our Tune" and knowing how difficult it could be to tune the little old radio, he would often switch it to different stations just to wind me up. It worked every time.

Rick and I were going out on some funerals, but not as regularly as we would have liked – something which irked both of us – but the volume of funerals was such that there was more than enough work behind the scenes for us to do as it was. My months of work experience had stood me in good stead as far as being able to fit and line coffins was concerned, but it was Rick who would induct me into the art of washing a car – properly.

There were many hints and tips for all manner of things which Rick had picked up from his time with his first employers and I learnt a great deal from him in those early months. Those first two years were for me all about learning the "nuts and bolts" basic tasks and skills involved in funeral directing and I was very fortunate to have such a thorough grounding. It must be remembered though, that all the time I was learning the basics I was also being immersed in the whole environment of funeral directing and this too would prove to be a valuable foundation.

In addition to car cleaning, coffin fitting, removals and a myriad other tasks, Rick and I were also responsible for all the mortuary work. I had been shown how to dress a body during my work experience, but with Rick and I now left to get on with these tasks ourselves we began to develop our own way of doing things.

The vast majority of bodies were dressed in a gown - a three piece garment with two sleeves and the main garment itself, that laid over the body and was then tucked down the sides inside the coffin, enabling one member of staff to be able to dress any size of body. The gowns always came as an interior set, with matching "curtains" to place around the head to cover the pillow, together with a frill that would then be stapled around the rim of the open coffin to complete the effect. Back then the usual colours the firm kept in stock were white, bright pink, the equally vile powder blue and oyster or champagne. Nowadays there is a far greater range of gowns available, in considerably more subtle colours and/or different patterned fabrics, but ironically the vast majority of families now opt for the deceased to be dressed in their own clothes anyway.

It was left to Rick's and my discretion as to which colour gown should be used with each body. The white was of course the safest choice, the blue, oyster and champagne sometimes requiring a little more discernment; but the pink was the colour that could really divide opinion. Some families would leave the chapel of rest saying how lovely Mum or Grandma looked, but on one occasion a very camp gentleman burst out of the chapel exclaiming that we would "simply have to change that ghastly gown. Mother would never have been seen dead in that shade of pink!"

Back then probably only a quarter of families requested for the deceased to be dressed in their own clothes and whilst gowns were designed with the ease of the person dressing the body in mind, with "own clothes" there was no such luxury. There is a persistent idea in people's minds that we just cut the clothing up the back and lay it on, much as a gown is designed to be used. However, quite apart from the fact that professional pride meant we regarded cutting only as an absolute last resort, it didn't actually make life a great deal easier anyway. So Rick and I, with varying degrees of ease, would wrestle with all manner of different clothing, ranging from the deceased's favourite casual, gardening or golfing clothes through to formal attire such as uniforms, dinner jackets and our chief nemesis, the three-piece suit.

On three very poignant occasions I have dressed young women in their wedding dresses, including a teenage girl who was buried in the bridesmaid dress she was robbed of the chance to wear, after losing her battle with terminal illness before the wedding took place.

For larger bodies especially, the scene would normally consist of me standing or squatting on the mortuary slab, supporting the body in a sitting position, whilst Rick would wriggle the clothes into place. Nowadays the process is far less physical, as, after years of practice, my technique has developed to the extent that the dressing of virtually any size body can be achieved single-handedly, with the body remaining horizontal and without any cutting of clothing. There's simply a knack to it, like everything else.

Only very recently a bereaved client told me of his experiences when dressing his father after he'd died. My client had discovered for himself just what a trial it can be without knowing the proper techniques. "Like a Monty Python sketch" were his exact words and at the reference to comedy sketches I was reminded of one particular, but for my part wholly unintentional, incident which caused days of hilarity to my boss Paul. During my early training Paul happened to walk into our mortuary just as I was trying to fit a bra to the body of a very large lady lying on the slab. He referred to my hamfisted handling of her very pendulous breasts – lolling about upon her horizontal frame – as "like watching someone trying to use one hand to catch two jellies and put them into moulds."

In addition to how the body is to be clothed, jewellery is always a matter of paramount concern to us as funeral directors. If the body is received wearing any jewellery, every item must be carefully recorded and the next of kin's instructions sought as to whether the jewellery remains in situ for the funeral or is to be removed and returned to the family.

Obviously, the utmost care must be taken at all times with jewellery, so imagine my horror when during those early years, I removed a ring from a body only for it to slip from my fingers, bounce on the tiled mortuary floor and dive straight into the drain. In utter panic I flung the little grate off the drain and with no thought for what might be lurking down below, I thrust my hand in. Crouched on the floor, with my arm submerged up to my elbow, my fingers firstly found the base of the drain and then touched the sunken ring. I didn't care that I was elbow-deep in the mortuary drain - I had snatched victory from the jaws of disaster and that was all that mattered.

I have explained about how we dressed and coffined our deceased clients, but I ought perhaps to explain a little more about how their bodies would be collected in the first place. I learnt very quickly that "removal" was the term applied to the collection of bodies. For someone in the office to finish a phone call and announce to anyone in earshot that "We've got a removal" meant by definition that another funeral order had come in, after which two staff would be dispatched in the removal vehicle. A "coroner's" or "police removal" as they were referred to back then meant that the coroner or the police wanted us to attend at a scene of sudden death as soon as possible and remove the body to the nearest public mortuary, although in such cases there was no guarantee that Thomas Broad & Son would eventually carry out the actual funeral.

Meanwhile, routine hospital removals had no immediate urgency and were carried out at our convenience, almost invariably by just one member of staff. On one occasion though, Rick and I were both sent on a routine removal at Cashes Green Hospital, a small geriatric hosital in the town. It was the same procedure at all small hospitals: park by the mortuary, wander down to the administration office, collect cremation forms when required and ask the administrator or receptionist to call for a porter to be sent "round to Rose Cottage". On this occasion a young and very camp porter was sent out. I can see him in my mind's eye now: grey warehouse coat, mullet haircut with bleached streaks and big glasses with bright yellow frames.

I must admit that Cashes Green was one hospital mortuary that I wasn't particularly fond of collecting from. There was an ante room where the porters kept the concealment trolley that they used for removing bodies from the wards, together with a hand washing sink and a desk for the mortuary register. Then, through another door was the body store itself. The room was set out as a chapel of rest with a large purple curtain covering one wall and a small altar near to the door, which always made it tight to get our stretcher trolley in. Concealed behind the large curtain was a three- body fridge, which, unlike every other mortuary fridge I had encountered, was a side-loader. This meant opening two huge insulated doors and then being faced with the bodies laying sideways on to you. I found that sideways format rather creepy - it always made me feel as if the bodies might suddenly reach out to grab me. It was simply because I was used to standard mortuary fridges where the bodies are inserted either feet or head first, so that even the most determined zombie would have had great trouble clambering out! What can I say? I've watched too many horror films...

However, on this occasion it was someone else's turn to feel uncomfortable. The camp porter, whose every word was causing Rick to fight the urge to snigger - and me with him, pulled back the curtain and opened the fridge door. He then turned and said,

"Ooh, I'm feeling fragile as it is. Would you two mind doing it all? I've just had my breakfast."

Rick and I were desperate to get the body out of there and be able to get back into the soundproof safety of the car so that we could finally let out the fit of giggles we had been holding in like a pair of naughty schoolboys.

Removals from private residences and nursing homes can be very problematic and the only way to train for this is by experience. Once I had started with the firm full-time, Paul and Michael finally took me out on house and nursing home removals. I had a lot to learn. Then, as now, we had different types of stretcher we could use and other pieces of equipment that could help, but ultimately it was down to technique and at times just doing what you had to. Houses are never built with undertakers in mind and neither are nursing homes. Obviously there was little we could do about quaint little cottages with winding staircases, but it was frustrating to have one particular nursing home that we regularly visited where, either by coincidence or plain bad luck, the body always seemed to be in a top floor room. Not only that, but the stairway was fitted with a stairlift. These wretched machines are the bane of every undertaker's life because they really get in the way when you're trying to carry a body down stairs.

On one particular removal from this nursing home, we were told that, yet again, the deceased was up on the top floor. Having made our way up to the room, Rick and I pulled back the bedclothes and weren't remotely surprised to find that the deceased lady in question was, yet again, of large proportions. Resigning ourselves to a heavy carry down the stairs we placed her on the stretcher and made our way down. We met the stairlift chair on the first floor landing and just about managed to negotiate that with our human cargo. But as we went down the final flight of stairs, squeezing past the chairlift rail as it was, I missed my footing as the stairs turned an angle. My weight pitched me over the stretcher in an almost head over heels roll and I was flung head first into the wall. I suppose it was fortunate that it was only nursing staff that saw this happen, rather than a bereaved family, but nevertheless a difficult removal can only get worse when you have a thumping headache and you're seeing stars.

Leaving aside the practicalities of removing bodies from varying locations, there was a far greater lesson that I was to learn from going out on removals from client's homes. I make no secret of the fact that I was brought up in comfortable, middle class environment and my life experiences up till leaving school had naturally reflected that fact. But death is no respecter of class or social standing and whether prince or pauper, or anyone in between, we will all die. Consequently, as undertakers, we deal with bereaved people from every conceivable point on the social spectrum. My innocence was to be truly challenged in those early years, going to all kinds of different households in all kinds of areas. My early thoughts and reactions seem excruciatingly naive to me as I recall them now, but I can only record how I felt and thought at the time, bearing in mind that I was just 16 years old, had just left school and had no experience of the world I discovered beyond my own sheltered existence.

I vividly recall one removal, with Rick, where we were sent to the terraced house of an elderly man who lived on his own. The dirt on the outermost of the three pairs of socks the dead man was wearing – and which he hadn't taken off for a very long time – was so encrusted that it was actually starting to buff up and shine from the friction with his trousers. There was a biscuit tin by his chair which he had been using as a toilet and the whole house was filthy and untidy. After the removal I congratulated myself on coping so well with such squalid conditions. Squalid?! That was to be nothing – barely a blip on the scale. In time to come I would learn what real squalor looked and smelt like.

The most ridiculous episode occurred late one gloomy winter afternoon, when I accompanied my boss Paul to a removal from a house on what was admittedly a rather ripe council estate on the outskirts of the town. I waited in the car as usual whilst Paul went in and had a preliminary chat with the family. Upon his return to the car he opened the driver's door, slung his clipboard back onto the seat and said quietly "Come on then old bean, lets get on with it."

As Paul opened the tailgate and pulled out the stretcher trolley, a hard-faced man in his thirties came down the garden path wielding a garden spade. Without uttering a word he swung the garden gate right back on its hinges and in one swing of his other arm he thrust the spade downwards, letting go of it just before it stabbed itself into the turf in front of the gate. "Alright then? You need a hand with anything else?" He asked, looking at Paul.

"No that's fine thanks. We can take things from here." Paul replied in a very relaxed tone of voice.

The man turned and stalked back into the house again. Meanwhile, I was still rooted to the spot, having been left fearing for my life before realising with huge relief that the spade was only going to be used to hold the garden gate open for us. I had so very much to learn.

The first act on a removal is to check the body for jewellery. Likewise if the deceased is dressed we will always check the pockets for any personal items as well. I was out on a late night removal, from an address in the hillside village of Amberley, a solidly Conservative-voting little enclave where its leafy seclusion is reflected in the house prices, when another item the deceased was wearing escaped my attention – at first.

As Len, one of the part-timers, and myself were travelling back to the funeral home with the body we thought we could hear a soft, ghostly sound, like singing, coming from the back of the removal vehicle. We listened carefully and sure enough there was the haunting melody again, emanating from under the cover on the stretcher. When Len and I got back to base it was with some trepidation that I took the cover off the trolley and unzipped the flexible stretcher to expose the body. As I started to unzip the singing noise started again.

It's not unusual for a body to make a very disconcerting croaking noise when it is handled – this is caused by air escaping from the lungs when disturbed by the movement of the body, but this time it was definitely a singing-type noise. We uncovered the body and firstly saw to our relief that it was still very much dead. Then I realized where the sound was coming from: the deceased was still wearing his hearing aid and the ghostly melody was nothing more than feedback noise from where the hearing aid had been enclosed inside the sheet we had used to wrap the body!

I was fortunate to start work at a time when the original combination of builder/undertakers were still trading locally. They were a part of my new profession's heritage and I've always been grateful to have witnessed this little bit of living history first hand. In all there were still eleven funeral directors trading in the Stroud district when I started work. My employers were by far the largest and the longest established and their only real competitors were the two other full-time companies, each based four miles away east and west, in the small district towns of Stonehouse and Nailsworth respectively. With the exception of an almost dormant branch of a well-known funeral directing chain, whose premises were later to figure so largely in my future, the other funeral directors were all local builders or hardware shop owners whose principals were still carrying on the long tradition of performing funerals as a sideline – in some cases just a handful of funerals a year. Thomas Broad & Son supplied vehicles and in some cases coffins, to five of these companies.

One of these builder/undertakers, Lewis Blackford, would arrive in his old Volvo estate car – finished in a very unfunereal custard yellow - to collect coffins from us, invariably taking the opportunity to bore me rigid with stories of how much better coffins were when he used to hand-make them. *Why did the daft old bugger come and buy them off us now, then?* I used to think. We would also then send the hearse and requisite number of limousines to Lewis' premises on the day of the funeral. His office was a quaint little red brick building with a tiny, neat garden in front and was located by the roadside on the far side of a hump-backed canal bridge just behind my old school. His little premises consisted, so far as I could see, of just two small rooms - the office for his building company and his Chapel Of Rest. His building yard itself was a few yards further down the road.

I was sent down to Lewis Blackford's on one occasion to act as a bearer for the funeral of a stillborn baby. There was a considerable quantity of flowers and so, in place of the customary, suitably-prepared estate car which would normally have been used for an infant's funeral, the hearse was booked instead. Myself and the hearse driver had placed the tiny white coffin in the vehicle and then arranged all the floral tributes around it. Despite initial misgivings about using a full-size hearse for such a tiny coffin, the presence of all the flowers did make the tiny coffin look a little less incongruous.

As we sat in the hearse waiting for Lewis Blackford to appear out of his office, I looked back at the hump-backed bridge and was suddenly reminded of being back at school. One of our more popular teachers, when taking us out on occasional study visits, would always drive the school minibus really fast over the small bridge so that we would all get that "leaving your stomach behind" sensation as the minibus landed on the other side. The school had a couple of minibuses, but the one that we all enjoyed riding in the most had bench seats running down either side and in those pre-seat belt days this teacher would always take bends really sharply just to throw us round inside the minibus, with all of us in the back allowing ourselves to end up in an unruly pile of bodies in one side or other of the minibus – much to our delight of course.

Although I had never really enjoyed school, in that moment I suddenly realised why people always said that "schooldays are the best days of your life." Being at work sometimes felt like being at school again, because being the junior member of staff I was always told what to do or what *not to do* and I would sometimes get really fed up with having Paul or Michael on my case, as if I hadn't already learnt the standards that staff were expected to maintain. On more than a few occasions it would be a classic case of "do what we say, not what we do" and it rankled with me. When it came to learning how to bite my tongue, I could have gone for olympic gold.

One of the other builder/undertakers always required us to help with the initial removal of the body each time, as well as having vehicles and a coffin supplied. He had a very old-fashioned workshop with a tiny yard adjoining it, at the top of the town. His chapel of rest was very utilitarian and bereaved relatives would have to walk through the workshop itself just to access the chapel. This used to horrify me, as in our own firm I would sometimes get told off just for something as small as opening the door from the back corridor too loudly if I happened to come through when there was a family in our chapel. The builder's unpolished approach didn't stop there either. Unlike all his contemporaries around the district who would all wear either a tail coat or frock coat when conducting funerals, this builder would always wear a black anorak.

Once I was able to drive I would take my turn in being sent up to his yard to go out and perform removals with him. One particular removal, just a street away from his workshop, presented us with an unusual difficulty. From the outside, the house seemed as unremarkable and as tatty as its neighbouring properties. However, the moment we entered – no, squeezed - through the front door, we discovered the deceased was an obsessive hoarder. Bundles of newspapers were stacked floor to ceiling to the extent that the occupant had actually created a warren of tiny walkways through all the bundles just to get to each part of his house. Mercifully I've never suffered with claustrophobia – just as well as we literally had to breathe in just to squeeze down these little walkways. Clambering up the stairs to the bedroom where his body lay was an expedition in itself. The dead man was led in bed, surrounded of course by all manner of junk.

All the while we had been edging our way through the house I had been trying to work out just how we could extricate the body from the property. In a normal house removal situation we would take the rigid trolley stretcher as close as possible and then use a flexible stretcher to carry the body from the bedroom and down the stairs. The issue here was that we could barely squeeze our upright, living bodies through this mini labyrinth, let alone a dead body on a stretcher. We identified a slightly easier route to the back door, but after wrestling with this door we found the garden was also waist high with hoarded building materials. We had no option but to clear a path through the house as best we could, sacrifice any attempt at dignity and just blunder on through. We eventually managed to remove the body, with a lot of heaving and huffing, a bit of quiet swearing and some grazed knuckles. Mercifully however, there were no relatives present, just the neighbour who had let us in and then wisely retreated back to his home next door.

My first years with my employers were to witness all of the builder/undertakers winding up, with just one doggedly carrying on - much to the irritation of his sons who, also being co-directors in the building company, were repulsed by anything connected with the funeral side of the business. They tolerated the funerals whilst their father was still there to deal with them, but upon his retirement they immediately converted the chapel of rest into much-needed additional workshop space. However, before their father retired I had become self-employed and I would end up hiring my own vehicles to his company – an arrangement from which arose an interesting experience I will tell of in a later chapter.

Back at my employers, Thomas Broad & Son, when a family requested a solid timber coffin – solid oak being the most commonly requested - or if we just needed an unusual size, then we would order it individually, as a "special."

My first trip to our main supplier to collect a "special" was quite an eye-opener. It just so happened that one of the largest coffin manufacturers in the country was located about eighteen miles away from us and the first time I was sent up there I was shown around by the factory manager. I was fascinated to see the bewildering array of differently styled coffins, all in neat stacks in every spare corner of the factory floor. I tried to imagine where each of those coffins was destined for and what stories lay behind the choice of some of the more ornate or unusual styles.

Watching queues of coffins trundling along a huge conveyor that threaded its way round the factory floor was a bizarre sight. On a later trip to collect another special, the guy that was dealing with me told me about his progression through the factory to his current rank of supervisor. Apparently his was a meteoric rise, starting in "stores" before cutting his teeth in "lids and bottoms", extending his skills in "mouldings" and the "spray shop" and when they were really busy he might also lend a hand in the "fitting shop", where, with the help of pneumatically-powered hand tools, they could fit and line coffins in a third of the time it took me. The person at the factory I always felt sorry for was the one poor guy, with a perennial "anywhere-but-here" expression on his face, stationed right at the end of the conveyor belt in the loading and collection area, screwing down lid after lid after lid, ready for the coffins to be stacked to await the delivery lorry. I often wondered if he ever got to do anything else in the factory.

We would take stock deliveries of standard style coffins once a month usually. It was a very physical task unloading forty coffins at a time and so it was always with a sinking heart that I would hear the sound of the lorry pulling into the yard. Not only that, but to my intense irritation the lorry often arrived during lunchtime. Michael would come thumping up the stairs to the first floor rest room that we had at the funeral home and announce with barely concealed, childish delight that I was needed down in the yard to unload the coffin delivery. Lunchtimes were one of my few little luxuries and when he disturbed even those times I always felt like I wanted to flatten him.

One particular day I happened to be standing at the top of the yard when I spotted the coffin lorry trundling up the steep drive. As it made its way up I could hear a faint rumbling sound above the noise of the lorry's engine. I had a sudden and horrible feeling of forboding. The delivery driver parked up by the main coffin store (located across the yard from the coffin workshop, where I only had room for half our normal stock), clambered out of his cab and opened up the tailgate.

The driver and I stood there in silence for a moment, surveying the scene of devastation in the back of the lorry. I suspected what had happened as soon as I heard the rumbling noise when the lorry was climbing the drive. Our coffins were all in a jumbled pile in the back of the lorry, lids askew, decorative mouldings sprung and snapped, sides scratched and corners chipped.

The driver broke the stunned silence.

"The bloody load strap's snapped. Must've been the weight of all the coffins pushing against it when I came up the drive."

"Well," I said, "It was just a routine stock order. We're not *too* desperate for coffins at the moment, so let's unload the survivors and then we can toss a coin to decide which one of us will have to go into the office and break the news to my boss."

One particular aspect of coffin preparation I had to learn once I started full-time was the furnishing of children's coffins and baby caskets. The majority of babies that Thomas Broad & Son dealt with had died in the county's maternity hospitals and as an integral part of the hospital protocols the bereaved parents would be offered a "hospital funeral," whereby a funeral director contracted to the Hospital Trust would arrange a very simple burial or cremation, at which the hospital chaplain would officiate. Alternatively, the parents could choose to appoint their own funeral director and make private arrangements themselves.

Unlike nowadays, where at least half of the baby funerals my company arranges involve pre-term foetuses, during the first years of my career I would only ever encounter full term or slightly premature babies, who in turn were either stillborn or who only lived for a short time. The company always kept a selection of different sized, rectangular, plain chipboard "shells" in store and when a baby funeral came in I would then prepare the appropriate shell by covering it in domette - an embossed white fabric designed for lining coffins. We always kept a stock of miniature coffin handles and small nameplates to furnish these coffins with. The fittings for baby coffins were always nickel plated – nickel being referred to as "white" metal (whereas brass plated would be referred to as "yellow metal"). The white coffin, with its white metal furnishings was supposed to be symbolic of the purity of childhood.

Nowadays the trend towards coffins made from natural materials has found a very useful outlet with children's coffins. There is something really very appropriate in using woven willow or bamboo for baby coffins because they have a much softer aesthetic than traditional wood. Another policy my company has adopted nowadays is always to ensure that children's coffins are ordered slightly larger than is necessary, in response to lessons I learnt whilst with Thomas Broad & Son. Bereaved parents invariably want to place toys in the coffin with their child and one child I remember very clearly, a seven year old boy, was to be buried with so many of his toys that Rick and I had to spend considerable time carefully re-arranging all of the toys inside the coffin just so that we could fit the lid on. I always remember Rick saying:

"We'd have been better off putting him in the smallest sized adult coffin – at least he'd have had room for all his toys then." It wasn't a callous remark.

The view in the firm was always that children's funerals required particular thought and sensitivity and that simply miniaturizing everything we did for adult funerals would have been both lazy and lacking in thought. So the general rule we adopted was that if the coffin was three feet or less in length, then we would use a rectangular casket shape, but over three feet long and the traditional tapered coffin shape would be used, subject always of course to any wishes the parents may have had on the matter. Again, subject to the parents' wishes, the coffin would then usually be transported in a suitably prepared estate car rather than a hearse, although some parents chose to travel in a limousine with their child's coffin on the seat between them – an equally kind solution, I feel.

I quickly found that, contrary to popular belief, a child's funeral does not cast a pall of gloom over the funeral director's office and neither did any of my colleagues or seniors ever find them unbearably emotional. What in fact makes children's funerals a challenge is in having to be sensitive to things that might not even be considered in the case of an adult funeral. It needs a great amount of care to identify these things and then find an appropriate response.

Although I'm not a parent and thus have nothing by which to measure my understanding of the subject, the one thing that I do always find incredibly poignant is having to place toys in those tiny coffins. I often find myself imagining the happy anticipation with which the parents-to-be would have bought those toys, before suddenly having their hopes and dreams cruelly snatched away from them and being left to give those toys to a child who would never be able to play with them.

However, having said that I do feel really uncomfortable when funeral directors or other death professionals encourage the notion that "children's funerals are the hardest." Of course that's true to an extent, but I think that by reinforcing this idea they are, in effect, trying to *quantify* the level of tragedy. In a sense they're saying that one type of bereavement is automatically more tragic than another. To me, that kind of thinking is extremely dangerous for anyone professionally involved with the bereaved. Let me illustrate my point:

After the first year or so, I was trusted to oversee a good number of the out of hours "viewings" (the chapel of rest visits by families). I was living close by and I was keen to be involved with as many different aspects of the work as possible. I vividly remember one week when there was a viewing for a stillborn baby on the Tuesday evening and a viewing for a lady in her nineties the following night. The Tuesday evening viewing I was expecting to be very emotional and probably quite long in duration, but the elderly lady was, I assumed, not likely to be a great concern because the family would probably be in and out within ten minutes, like relatives usually were in such cases.

The baby viewing took about twenty five minutes and all I could hear from the chapel was silence. The young parents came out of the chapel understandably choked with emotion, but nevertheless they offered a barely audible "thank you" as their faces quivered uncontrollably with held back tears. I stood there feeling rather useless.

The elderly lady's viewing however, was to prove altogether more dramatic and instructive than I imagined. The lady's two sons, together with their wives and a teenage grand-daughter all went in together. After quarter of an hour, during which I could hear the faint noise of talking from the chapel, the grand-daughter started getting louder and clearly more emotional. The family remained for another five minutes, during which it sounded as if the girl had calmed down again. The chapel door opened and the four adults stepped out and then paused, waiting for the grand-daughter to follow them. But as the girl reached the chapel doorway she suddenly tensed, gripped the doorframe and refused to move, saying firmly that she "didn't want to leave Grandma." Her father took her by the arm and managed to gently prise her hands off the door frame but as he led his daughter across the hallway to the front door, she tensed again, began screaming for her grandma and then dug her heels into the carpet. With the two men of the family there anyway, I decided it was best for me to just stay out of it. After the girl's parents' attempts to calm her failed, the hysterical teenager was manhandled out of the door and across the car park, all the while with her heels stuck firmly into the ground, to the extent she left two track marks across the gravel to the car. All the while she was still screaming for her grandma.

That was my first encounter with violently expressed emotion and the experience left me rather shaken at first. To that teenage girl the loss of her ninety year old grandmother was every bit as acutely painful as the loss of their baby son was to the young couple the night before. The lesson I learnt that second night, although I was still to see many more examples to reinforce my learning, was that any bereavement can be a profound personal tragedy for those concerned and it is very dangerous to judge the level of someone's loss just by the outward circumstances.

CHAPTER FIVE

Smoke And Mirrors

"Nothing is more detestable to the physical anthropologist than... [the] wretched habit of cremating the dead. It involves not only a prodigal waste of costly fuel and excellent fertilizer, but also the complete destruction of physical historical data..."

Up From the Ape -1931 (*Albert Ernest Hooton*)

"I always thought they just scooped it all up and gave you a bit." A friend declared, when describing her perceptions about cremation. She went on to say:

"That's the main reason I didn't collect my Grandmother's ashes afterwards. I thought I'd just end up scattering Tom, Dick and a little bit of Margaret."

The cremation process is so bound up in modern myth and misunderstanding that I couldn't even begin to count the number of times that I've been asked if we "re-use the coffin" or if it "really is the right person's ashes that you get back after cremation." I've always answered a firm "no" to the first question and an equally emphatic "yes" to the second, before giving a brief explanation as to how the entire process of cremation is subject to strict regulations enshrined in law and how an unbroken chain of identification is maintained throughout the whole procedure. I have of course witnessed it all myself.

During a quiet week early in my new employment, my boss Paul kindly arranged for me to spend a day with the staff at Gloucester Crematorium and gain an insight into this aspect of funeral work.

Looking back, I'm glad I first experienced the workings of a crematorium before computer technology took over, when everything was still manual and mechanical in an almost quaint kind of way; when playing recorded music meant cueing cassette tapes to the right piece of music and the responsibility of ensuring that up to thirteen funeral services a day ran smoothly was the duty of the chapel attendant - that most discreet, but vital, human cog in the wheel.

Gloucester Crematorium is located in the Coney Hill area of the city, an address branded with an aura of deterrence by the presence back then of the old Coney Hill Mental Hospital. Originally built in the Victorian era as the City & County Lunatic Asylum, the concealing isolation of its large grounds only compounded the fearful ignorance, the lurid folklore and the defamatory language that local people imposed upon the institution and its inhabitants.

Historically, local prejudice had always reasoned that the expansive grounds, over which the hospital clock tower stood as a grim landmark of lunacy, were necessary to prevent the escape of marauding packs of dangerously insane inmates. Just the fact patients were described as being "committed" to the hospital perpetuated an idea in people's minds that even the most tame and mentally-improved patients were still regarded as inmates to be kept locked away. It was inevitable then, that just the mention of "Coney Hill" conjured up in local people's minds the dark silhouette of isolation and confinement, replete with equally dread, if very outdated, ideas of certified and detained lunatics, straitjackets, padded cells, white coats, derangement and dangerousness. Post-war residents of Coney Hill must have wondered why fate chose to deal them a cruel hand when, during the 1950's, they found themselves as neighbours not just to the county mental hospital but then also a crematorium too.

I rode over in the hearse on an early funeral of the day and during the service I was introduced and handed over to Patrick, the crematorium chapel attendant. Paul always spoke of Patrick in very warm terms and described his reputation among the funeral directors as the stuff of local legend. A very quiet, self-effacing man, Patrick, I was told, ruled the chapel brilliantly, keeping everyone and everything under constant control and ensuring that no matter how large or small each funeral was, that everything was kept running to time and in an orderly fashion – at times a far greater skill than might be imagined.

It's that constant pressure of time that colours the public's perceptions of crematoria and if I'd been given a pound for every time someone has said to me over the years that "the crematorium is just like a conveyor belt" I wouldn't need to be working now. Normally I try not to get too bogged down in that discussion, but regardless of whether the remark is meant as an accusation or an observation, the challenges that face crematorium staff and funeral directors deserve some explanation:

Roughly 70% of funerals in the UK involve cremation (another question people often ask). Needless to say, that creates a constant demand on crematoria around the country, but nevertheless every individual crematorium still requires a certain volume of cremations per year in order to remain economic. Indeed there has never been a more expensive time to establish a crematorium, particularly with the cost of compliance with EU regulations on environmental emissions - fumes, to you and me.

With so many families wanting to use one facility in any given area, the challenge for crematorium staff, funeral directors, ministers of religion and secular officiants, is to make best use of the time available. We have to be able to marshal anywhere from a dozen mourners up to hundred or more without making them feeling like herded sheep; plan for a meaningful ceremony to take place within a limited time span and also deal with those dreaded times when a previous service has overrun, or a cortege is delayed en route to the crematorium. I always say, that as a funeral director, I have a responsibility not only to my mourners, but also to the funeral director and mourners on the following funeral. It's a very delicate balance, requiring effort on all sides. A crematorium can build an additional chapel, extend its service times from say 30 minutes to 45 minutes, or provide separate entrances and exits to ensure different sets of mourners do not have to bump into each other, but in the same way as building extra roads does little to reduce traffic jams, so too at crematoria the fundamental problem of demand for the service always remains.

But returning to my day at the crematorium, Patrick, Paul and myself slipped quietly back inside the chapel and sat at the rear to watch the rest of the funeral. As the curtain closed across the coffin – the part of the service referred to as the "committal," Paul informed me in a whisper – I looked towards the main family mourners and I could see them shudder with emotion.

With the benefit of many years experience, I understand now just how final that moment is for the bereaved. With a burial, the coffin is still in sight at the bottom of the grave and ultimately the family can choose when to walk away from that open hole in the ground. But there is a very cold, final and disenfranchising quality about watching the curtain close at a crematorium. The act of lowering a coffin into a grave is a simple, natural concept anyone can understand and relate to, whereas cremation is often overlaid with mystery and mechanical artificiality.

After the service drew to a close and Paul had led the mourners out of the chapel, Patrick took me through a side door behind the minister's lectern. As he led me through he explained I would spend the morning in the company of Ted and Roy, the two men who between them operated the crematory itself and supervised the music room. The moment didn't have that air of nervous anticipation that my first experiences with bodies and mortuaries had, because this time I would only see coffins and cremated remains, but nevertheless this was still going to be another step into the fascinating unknown for me.

Patrick first introduced me to Roy, who was sat in the tiny little music room adjacent to the chapel. Nowadays, with all the push-button technology at crematoria, it seems incredible to think of Roy staying couped up in that little grey room with its tiny windows high up in the wall, waiting for the minister conducting each service to press buttons on the lectern that would illuminate a series of coloured lamps on the music room wall. Each different coloured lamp gave a signal to Roy as to whether to play the recorded music or close the curtains during the words of the committal. In addition to operating the cassette player (as it was then) Roy had to operate the curtains in the chapel by means of a hand cranked pulley mounted on the wall in the music room. There was no means for the minister to communicate with Roy other than by the coloured lights, so a running order always had to be given to Roy beforehand -especially for funerals requiring the playing of more than one piece of music. He seemed to have a permanent nervous smile and I rather unkindly wondered whether he was on day release from the county mental hospital less than half a mile up the road.

After letting me spend the duration of the next funeral observing how the music room was controlled, Patrick reappeared and took me through to a hallway behind the music room, where two of the bearers from the funeral that had just taken place were waiting. Patrick opened another door, beckoning myself and the bearers through. The four of us were gathered in a long, corridor-like area.

"This," Patrick explained to me, "is the transfer chamber." He had raised his voice slightly to make himself audible as it was surprisingly noisy out there and everything suddenly felt very industrial. To the left was a set of grey double doors, with a green light illuminated above them. On the opposite wall were three metal hatches. Patrick opened the double doors and there, just behind, was the coffin, resting on the catafalque where it had been placed at the beginning of the funeral and now hidden from view by the closed curtains. The green light was a signal that all the mourners had left the chapel and that the coffin could now be removed. Patrick pulled the coffin through onto a roller-topped trolley and let the two bearers step forward and take the flowers off the coffin lid before leaving the chamber.

"Once the coffin is placed on the catafalque in the chapel, the funeral director's responsibility for that coffin ends and the crematorium's responsibility begins" Patrick informed me. "Once that coffin has been placed on the catafalque you are allowed to do nothing more than remove the flowers from it." His tone of voice had changed and I could tell he was trying to impress upon me the importance of the regulations that bound both the crematorium and the funeral directors that used it.

Another man, in a grey warehouse coat had appeared from somewhere behind me. Patrick introduced him as Ted, the cremator technician. Ted uttered a brief "hello" and carried on with lining up the coffin-laden trolley with the middle one of the metal hatches on the opposite wall. Patrick continued his commentary:

"As I expect Paul has explained to you, the entire coffin has to be made of combustible material, as it's cremated with the body. We don't remove handles, we don't remove the body, all we do is check that the nameplate has the correct name and then we *charge* it straight into the next available cremator."

With that Patrick checked the coffin nameplate against a small card he was holding. Ted, meanwhile, had been turning a hand-wheel on the wall that opened the metal hatch and once he too had compared the card to the name plate on the coffin, Ted gave the coffin an almighty shove, sending it scooting along the trolley rollers and into the fire-brick interior of the cremator.

Having got used to seeing how gently and reverently the coffin was handled on funerals I was really taken aback at how violently Ted had pushed the coffin into the cremator. However, the coffin quickly slid to a halt just inside the cremator and although there were no jets of flame visible, I could still feel a blast of heat from the open hatch as I watched Ted try to shove the coffin, by now sat firm on the fire-brick hearth, an inch or two further in with his hand. I realised then why he'd given the coffin that rather undignified and almighty shove in the first place. He wound the hand-wheel again to close the hatch, took the small identity card from Patrick's hand and walked back through the doorway from where he had appeared. I happened to notice that next to each cremator hatch a square metal fire shield was hung on the wall. I tried to imagine what kind of circumstances would require Patrick or Ted to resort to using the shields. Considering the sheer heat I'd felt when Ted charged the coffin into the dormant cremator, I thought it would take a very brave soul to try putting a shield over the hatch during a mid-cremation emergency, rather than just making a run for it instead!

"Ok James, follow Ted round into the crematory and he'll look after you while I go and prepare the chapel for the next service."

The crematory was a very noisy, completely industrial-looking environment, with three huge cremators roaring away. It all felt like an old fashioned factory or a foundry might have done and I felt as if I had suddenly entered into a scene from one of those L.S. Lowry paintings, with the smoking factories and his famous "matchstick men."

I turned to watch what Ted was doing. He placed the identity card into a slot on the front of the middle cremator, the same one into which I had already watched him charge the coffin and he told me to look through the thick glass peep hole on the front of the cremator. Ted explained that this was how he observed the progress of each cremation. He told me to watch firstly how the veneer on the coffin would peel in the blast of the flame jets as he started the cremator up again. Then, prompting me to look through the peep hole of the adjacent cremator, he explained that this other cremation was halfway through. I could see the sides of the coffin, half burnt away and leaning at angles on either side of what could only have been the remains of the body. I could make out what looked like a pile of embers glowing in the jet blast, but as I continued staring through the peephole I began to make out the shape of the body. Ted explained that each cremation took an average of 90 minutes, with larger bodies cremating much quicker than small, thin ones – because the body fats actually helped the combustion process.

Ted pointed to the three cremators.

"The crematorium has capacity for up to 13 services each day, taking place at half hourly intervals and all those coffins have to be cremated the same day, so we've got three cremators to cope with the volume."

There was a temporary lull in activity as the three furnaces roared away at their work and Ted started fiddling with a broom handle to which he was trying to fit a new brush head. He asked me the usual questions about how I got interested in the funeral profession, making conversation while he put his new broom together.

After a few minutes of conversation, and with his broom successfully assembled, tested and commissioned, Ted led me into a small yard behind the crematory and pointed to a mirror mounted on a bracket up on the crematory wall. By using this mirror, Ted explained, he could see how much smoke was coming from the crematorium chimney and then by means of adjustments to valves on the cremator furnaces, he could ensure that unsightly plumes of dark smoke could be avoided. Although the monitoring mirror was laughably low tech, Ted demonstrated the huge difference a few adjustments to the cremator valves could make to the chimney's smoke emissions and he impressed upon me the importance of controlling the visual element of the smoke, not just for visiting mourners but for the residents of nearby homes. In later years, particularly driving limousines, I would get sharp-eyed mourners looking over my shoulder saying "Look, see the smoke. That's another one going up. Be Mum's turn in a minute" and each time I would remember Ted's explanation of the importance of monitoring the chimney.

From beyond the little wooden door in the yard wall I could hear the voices of mourners gathered outside, chatting after a service. Just the thickness of that door separated the displays of floral tributes and the quiet, well tended grounds from this noisy, industrial environment where I was watching Ted burn human bodies in huge, gas-fired furnaces. Once again that familiar sense of separation hit me: two parallel universes, physically situated cheek by jowl, but still light years apart. Meanwhile a third cremation had reached completion.

"When the last glowing ember has died, the cremation is complete." Ted intoned in a similar tone of voice to that which Patrick used when referring to the regulations. He left me wondering whether that was the actual wording of the regulations, because Ted just didn't seem the poetic type. He opened a small hatch next to the peep hole on the front of the cremator furnace and used a very long handled, square-ended rake to gather the disintegrated remains and pull them forward, allowing them to drop through a hole in the fire brick hearth down into a second furnace chamber below.

Although there were no burner jets in the lower chamber, by adjusting valves Ted could still direct further heat onto the remains to ensure that the last remnants of wood from the coffin were completely destroyed, so that all that was left were the cremated bone fragments. These in turn he raked forward again and collected into a large stainless steel tray which plugged into the front of the cremator.

I watched closely as he took the identity card from its slot on the front of the cremator and slid it into a similar slot on the lid of the steel tray. Other trays, each with an identity card attached, were already put aside whilst their contents cooled. Ted opened the lid one of the metal trays and showed me the contents. I could easily recognise some parts of the bone structure, like the hip joints and curved segments of skull, but they were all just broken, fragile fragments, interspersed with lots of distorted metal nails and screws used in the manufacture of the coffin. Ted was holding a peculiar, chunky little device in his hand that he began picking and grinding through the cooled-down set of remains.

"It's a very powerful magnet. We use it to pick through the remains and remove all the metal fixings from the coffin."

When the magnet was covered with metal bits he would hold it over a bucket, twist a small lever on the magnet handle which somehow deactivated it, allowing the bits of fixings to drop into the bucket. Then he would start the process again until all the metal was removed from the cremated remains. He showed me the contents of the bucket, which included the familiar spiked shape of a metal hip joint.

"All the metal we recover is buried in the cemetery grounds. We got a special hole with a locking cover and we tips all the metal collect-ings into there. When the hole's full the guys in the cemetery division take the cover away, top up the hole with earth and dig us a new hole. Precious metals is usually either vapourised or fused into unrecogni-sable material because the cremator is running at about eight hundred degrees. By burying all the metal material we recover there's no accu-sations about us re-selling precious metals or anything like that." Ted explained.

That all made eminently good sense to me.

"Now the remains have been sifted by magnet they have to be reduced to their final form" Ted announced, as he the carried the tray into a small room at the back of the crematory. He tipped the bone fragments into a round hatch on the front of a squat, square machine he called a "cremulator." Once again the identity card was taken from the tray and slotted onto the front of the cremulator.

Ted asked me to look inside.

"There's a rotating drum there and them three heavy iron balls you can see will reduce the bone fragments down, so they end up like sand granules. You've seen ashes before, have you?"

"Oh yes," I replied. I was genuinely interested to see how the many sets of ashes I had seen before at the funeral home actually got to that granulated form they were in when I poured them into scattering urns or burial caskets. The room where the cremulator was situated had a wall full of shelves, each shelf marked with the names of the different funeral firms that used the crematorium regularly and each shelf was stacked with varying numbers of urns and caskets waiting for collection. Seeing the shelf marked Thomas Broad & Son I instantly recognized those familiar burgundy plastic sweet jar-shaped urns that I had seen dotted all round the funeral home in Stroud. "Polytainers," Paul and Michael referred to them as; another bit of terminology I had stored away in my head on hearing it.

After being totally absorbed watching and occasionally assisting whilst Ted juggled multiple cremations for a couple of hours, Patrick reappeared. It was a bit of a wrench to leave the crematory area. I suppose the novelty of being behind the scenes in a crematorium had a lot to do with it, but there was something really appealing about being in that Lowry-esque environment out back, its industrial feel given a strange fascination by virtue of the fact that they were burning dead bodies, rather than just producing factory products. Like everything else connected with handling the dead, there was a uniquely transcendental quality to it that made it feel very special.

Patrick wanted me out front of house in the chapel again, to observe how the chapel duties were co-ordinated during another three funerals. At first it seemed rather dull in comparison to the actual cremation process, but I actually found new interest just watching three other funeral directors at work, noticing the little differences in the way they did things in comparison to how I had seen my boss Paul conduct his funerals.

Our firm had the last booking of the day so I joined the men in the hearse for the journey home. They were using the Daimler, which had a bench seat in front big enough for three people. I found myself sandwiched between Paul at the wheel and a rather eccentric old vicar. Since leaving the crematorium we had been stuck behind an elderly mourner driving home from attending our funeral, but as we reached the edge of the city there was a two-lane section of road where Paul was able to overtake. The Daimler's huge engine gave an exciting, deep roar as we blasted past the elderly mourner's car.

"Hallelujah!" cried the elderly vicar, throwing his hands up in a gesture of delight. "Derek is a lovely chap but I really do think the Lord should think about encouraging him to give up driving now. He is rather past it at his age." The vicar's attention switched to the hearse. "This really is a wonderful vehicle isn't it? Tell me Paul, how many cylinders does the engine of this magnificent machine have...?"

That was all twenty five years ago. Nowadays all the workings in a crematorium are all push-button and computer operated. The music room is no more – replaced instead by a desktop music system at the rear of the chapel. The minister or officiant simply gives a verbal cue to the chapel attendant for the music to be played, provided either from CD's or even downloaded directly off the internet into the crematorium's music system. The hand-cranked curtain is also long gone – replaced with an electric operating device controlled by a button on the lectern.

Behind the scenes, the old fashioned, mechanical world Ted held sway over has long been replaced by a characterless, modern crematory - all clean, bright and technological. The furnaces themselves are all computer-managed for maximum efficiency, operated by computer consoles that wouldn't look out of place in Mission Control at NASA. Under current EU legislation, all UK crematoria now have to invest in vastly expensive mercury abatement filter systems (to reduce the airborne mercury arising from the cremation of the mercury in tooth fillings). The computers ensure the smoke and general emissions are filtered and reduced, so even the mirror on the wall has been made redundant.

However, the cremation process itself hasn't changed – Ted's successors still have to charge coffins, supervise the furnaces, extract the remains, filter them with the magnet and cremulate them. But whilst I'm sure the computerisation and emissions filtering make it all very efficient and good for the environment, the crematorium has lost all its morbid charm and mechanical simplicity. However, the one thing that computers can't improve upon, or replace, is the chapel attendant. A good attendant is still worth his or her weight in gold and whilst Patrick has long been absent and much lamented, cremated in his own workplace in the end, his professionalism lives on in many of his successors.

Looking further into the future, the march of progress may yet mean that cremation, and indeed burial, could be joined by a third and even a fourth way:

"Resomation," a concept created by a Scottish company, is envisaged as being housed in facilities exactly like a crematorium, up to the point at which the coffin is committed from view. However, after the curtain closes, instead of fire, resomation will use a water and alkali based method - also known as alkaline hydrolysis - to chemically break the body down. The process will normally take two to three hours - slightly longer than an average cremation. Once complete, a sterile liquid and skeletal remains will be left. The sterile liquid will be returned to the water cycle, whilst the bones will be reduced to granular remains and placed in an urn for scattering or interment, in exactly the same way as they are after cremation.

The company behind the concept claim that environmental research has shown that the substitution of Resomation for cremation as part of a funeral will reduce that funeral's emissions of greenhouse gases by approximately 35%, whilst the energy needed for the Resomation process in the form of electricity and gas is less than one-seventh of the energy required for a cremation. Likewise, they say that Resomation would produce no airborne emissions.

Alternatively, "Promession," originating from Sweden, would see the body being frozen in liquid nitrogen, which would make the body very fragile. It would then be vibrated, causing the body to break down into an organic powder. The powder would be introduced into a vacuum chamber where the water would be evaporated. The remains would then be ready for placing in a small biodegradeable coffin and buried in the living topsoil, where the coffin and its contents would turn into compost in about six to twelve months. The originators of Promession claim similar environmental benefits to Resomation, in that their process also requires much less energy than cremation and eliminates airborne emissions, although after Promession the final remains are more readily biodegradable than bone ash.

Resomation and Promession might arrive in the future, but in the meantime cremating the dead still has a twist in the tail because, despite appearances to the contrary, cremation can still create as many problems as it solves. Cremation is after all just a process; it's not an end in itself. When the body has been cremated, there are still the ashes to be disposed of. "Save the land for the living" was I believe, at one time, the motto for the Cremation Society of Great Britain and in this respect cremation is effective. However, even something as seemingly simple as the scattering of ashes can sometimes create unforeseen side-effects. Consider the situation that has evolved – quite literally - at the summit of Ben Nevis, the highest mountain in the British Isles.

Scientists have found that around the summit, scattered ashes have fertilised the soil and promoted plant growth on ground that has been barren for thousands of years. Minerals normally in short supply on the high ground are being imported within the deposits of human ashes, causing alarm in some quarters. The Mountaineering Council of Scotland for example, appealed to relatives to consider depositing their loved ones' ashes elsewhere, rather than scattering them on mountain summits, taking the view that when people visit mountains they expect to find a totally wild environment, not peaks where the natural ecology has been unwittingly altered by posthumous human fertilisation.

But if that sounds bizarre, then contrast it with a far more mundane reality at the other end of the scale, where lies the untold story of the thousands of uncollected sets of ashes languishing on shelves or in cupboards in every funeral director's premises up and down the country. Families, often with the best of intentions, can either never quite bring themselves to make arrangements for scattering or burial, or even simply forget the ashes are still there waiting. In my own company we have found that the sending of tactful reminder letters usually only has a limited effect on reducing the numbers of unclaimed ashes and it never eradicates the problem. In some cases we find that the next of kin have simply "gone away." There were literally dozens of urns & caskets stored in the cellar at Thomas Broad & Son's premises when I worked there and indeed all these years later I too have had to install sets of shelves in my own premises for exactly the same reason. Who would have thought that even happily married men and women would still finally end up "on the shelf" in death?

CHAPTER SIX

A Macabre Miscellany

"In all I saw eight post-mortems that first day. They were far too interesting to make me feel ill."
Evidence For The Crown - 1954 (Molly Lefebure)

From within the coffin workshop I had learnt to recognise the click of the swing door that separated the entrance hallway from the rear corridor, followed by the sound of a footstep on that one loose flagstone in the corridor that everyone stepped on as they passed through. These sounds were usually a signal that someone from the office was on their way and for me, as the most junior member of the firm, that usually meant trouble was coming. I picked up a screwdriver and pretended to be engrossed in fitting a handle to the coffin I was taking a break from working on, expecting my peace to be disturbed at any moment. Sure enough, the workshop door rolled back and in walked my boss Paul. As was his habit, he sunk his hands into his pockets and stood watching me for a moment, with a knowing smile across his face.

"I've arranged another day out for you my boy." He announced. "I've had a word with Professor Bassett at Gloucester Royal and arranged for you to spend a few hours in the mortuary. It might do you some good to see what goes on there and watch a post mortem being carried out. Those two 'erberts Ian and Gerry can show you what they do each day."

Ian and Gerry were the two Anatomical Pathology Technicians at the mortuary and although it was Ian who I'd met on my very first visit to the mortuary during my work experience, by that time I had also met his colleague Gerry many times too. He looked much older than Ian and for some reason he just seemed to fit in more with my idea of what a mortuary technician should be like. Gerry was half bald, sporting thick-rimmed glasses repaired with the obligatory bit of tape, he had a high-pitched voice and a way of speaking as if he were addressing you as a child. Gerry wasn't nearly as old as he looked – allegedly this was due to the effects of contracting tuberculosis from a post-mortem many years before.

In those early years the focus of my attention was very much on the mortuary side of things. At that very early stage of my working life I had nothing more than fleeting interaction with bereaved clients and my training was still very much concerned with "back of house" functions, of which there was still much to learn. Consequently I was delighted that I would finally get to see what the working life of the mortuary technicians was really like and of course finally be able to watch a post mortem.

A few days later Paul delivered me to the office of Professor Bassett – the head of the Pathology Department at Gloucestershire Royal Hospital. We had a brief chat before the Professor led me through the huge service corridor which ran from the main hospital, with one of its connections being the mortuary. The Professor summoned Ian out of the PM room and asked him to kit me out with a surgical gown, apron and gloves. Once I was gowned-up and we walked back through the body store area, a figure slumped in a wheelchair in a side room next to the PM room doors caught my eye. The side room itself was clearly a viewing area designed to allow authorised visitors to watch a post mortem without actually having to enter the PM room itself. As I looked at the figure more closely I could see it was actually only a home-made dummy, dressed in scrubs and a white coat, with a head fashioned from a white pillow and a Hitler caricature drawn onto it with black felt tip pen.

"That's Bert" said Ian with a wide grin, reaching his hand out to push open the PM room doors as he spoke.

I didn't get the chance to ask how, or indeed why, Bert the dummy had come to be there because Ian had already pushed open the double doors and finally the hidden mysteries of the PM room were set out in front of me. The first thing that struck me was how light and strangely friendly-looking the room was. The walls were a warm, peach colour and all down the opposite side of the room were a series of frosted windows, which bathed the room in sunlight. Ironically, now that I was finally stood in the PM room it actually felt a great deal less threatening than I had imagined it to be. A stainless steel work bench ran the length of the opposite wall below the windows and halfway along the bench I could see a pathologist stooped over a wooden cutting board, poring over what was clearly an internal organ of some kind. After waiting so long to see the sharp end of things – the really "up close, gruesome stuff" if you will, I was surprised by how unmoved I was by the sight of the dissected organ, instead feeling only a growing thrill at finally being here in the PM room.

Down the middle of the room stood a line of five stainless steel tables, shaped like inverted pyramids with grilled tops. Water was running from hoses connected to each of the tables, whilst above each of them electric sockets and operating theatre-style spotlights descended from the ceiling on metal columns like mechanical stalactites. Three of the tables were occupied by naked bodies, each with strips of paper towel placed over their genital areas. Gerry was leant over one of the bodies and he looked up at me. His familiar voice, muffled by his surgical mask, called out in welcome. From the look of his eyes, magnified by his big glasses, I could tell that behind his mask he was smiling broadly and it felt as if he was welcoming me anew, as if suddenly I was now a proper member of some sort of club.

This was 1987 - long before television programmes like "Silent Witness" or the interminable "C.S.I." series made mortuaries a trendy subject, so although I didn't know what to expect, at least my imagination was not coloured by those trendy, sexed-up television portrayals of PM rooms, with ridiculously dark, moody lighting and equally unrealistic hip, right-on technicians. There was nothing remotely hip or right-on about Ian and Gerry, despite Ian's unconvincing attempts to appear otherwise. The only television portrayal of mortuaries up to that point had been the hopelessly cheesy but still entertaining 1970's

US series "Quincy," about a crusading Los Angeles coroner's pathologist. Rick and I would often amuse ourselves by imagining Ian and Gerry being planted into the formulaic storyline of a typical episode of Quincy.

Dragging my thoughts away from an imaginary 1970's Los Angeles Coroner's Office and back to a very real 1980's Gloucestershire Royal Hospital, my eyes were inexorably drawn to the opened-up bodies on the tables. Up to that point in time my only experience of the physically disrupted dead had been seeing the body of a railway suicide victim whose funeral we had arranged and so I was expecting my intestinal fortitude to be severely tested, but I was surprised at how matter of fact everything felt to me. I can only think that continual exposure to bodies back at the funeral home had meant I was already starting to get used to such things without even realising it.

It was just as well that I felt like that, because post mortems are a uniquely practical form of surgery. Unlike surgeons operating on living patients, pathologists not only have the luxury of time, but they also have the luxury of being able to dismantle whatever they need to for closer examination. There was nothing remotely gratuitous about what I witnessed in the PM room, but having got used to being in there for that first time, I quickly began to feel like I was attending the ultimate in human biology lessons. I was able to see at first hand just what a breathtakingly clever bit of biological engineering the human body really is.

Gerry & Ian showed me how each body had been eviscerated (having the internal organs removed) before directing me to the dissecting bench where the pathologist demonstrated how he would then examine each organ in pursuit of medical clues as to the cause of death. The pathologist went on to explain that the brain also had to be examined and, as if mention of this was a cue, I turned to see Gerry pull a small motor unit on castors up to one of the tables. As he unwound the cable and plugged it in I could see in his other hand a small device containing a surgical saw with a crescent-shaped blade.

The pathologist led me back over to a body on one of the PM tables, explaining, in a rather "here's one I did earlier" kind of way, how Gerry had already made an incision from behind one ear, over the crown of the head, to a point behind the other ear and peeled ("reflected" was the term the pathologist used) the scalp away from the skull in two flaps to expose the skull cap itself. Gerry then ran the blade of the electric saw across the front of the skull and then round the rear, to create a kind of wedged "cap" that he then gently removed, exposing the membrane-enclosed brain. The pathologist demonstrated how the brain was observed in situ, before Gerry later worked to free up the connective tissues and sever the connection to the spinal cord, before lifting the brain out of the skull for detailed examination.

With that procedure over and once again taking in everything else around me, I felt an overwhelming sense of consummation in being in that whole environment; as if I was exactly where I should be. The most surreal thing about it all was that it didn't feel remotely surreal. I felt no physical repugnance or emotional response, but simply a sense of the utter normality in everything going on around me: watching the pathologist back at the dissection bench engrossed in examining the recently liberated brain, the rustle of Ian's plastic apron and the clopping of his hospital-issue clogs as he hovered between the tables and the dissecting bench, gathering up tissue and fluid samples in small containers and labelling them ready for sending to the pathology lab, the rasping hum of Gerry's saw on another skull – a sound disturbingly similar to that of a dentist's drill; and all of this to the accompaniment of Radio Two playing on the radio in the background.

I had crossed another threshold of experience, into another hidden world where once again the mystery of death was trumped by simple practicality; this time a place where everything was designed and equipped solely for the purpose of dissecting and examining dead bodies. But there was a far higher purpose behind this grim work: to provide peace of mind for the bereaved by pinpointing the reasons for how and why their loved one had died, whilst for medical staff came the benefit of further understanding diseases and conditions, yielding insights into how to refine treatments and hopefully go some way to preventing further patient deaths in the future.

I watched as Ian later began the process of reconstructing the first of the three bodies undergoing post mortem examination that day.

He carried the organs across from the dissecting bench in a stainless steel bowl, before gently placing them back in the open trunk cavity, packing absorbent tissue over them and sewing (suturing, to use the proper term) closed the trunk with that familiar zip-like understitching I was getting so used to seeing on coroner's cases brought back to Thomas Broad & Son's mortuary. Having sutured his way up the trunk of the body, chatting to me as he went, I now maintained a respectful silence as Ian stopped chatting, wrestling with what he briefly remarked was the most hated aspect of his working day: suturing up a scalp. He carefully positioned the sawn off cranial cap - the top half of the skull - and manipulated the loose folds of the scalp back into place over the now restored skull, before beginning the task of suturing the thick tissue of the scalp, all the while constantly combing the hair away from the incision to prevent it getting tangled in the suturing. It was the ultimate in invisible repairs and after he had finished the body once again looked whole and undisturbed, save for the tell-tale line of suturing from the throat down to the pubis, which itself would be hidden after the body was dressed.

By the time Ian had finished rinsing down and drying the newly restored body and I had helped him transfer it off the table and back out into the fridge, the clock was rapidly approaching 11.00am – opening time as far as funeral directors were concerned. Leaving Gerry to carry on in the PM room, we changed back out of our surgical gowns and whilst waiting for the arrival of the first of the day's funeral directors, Ian showed me round the rest of the mortuary.

He first showed me into his and Gerry's tiny little office. The term "office" was perhaps overstating things a little – "mess room" might have been a more appropriate term, with the accent on mess. Ignoring the chaos of paperwork and sandwich wrappers, my eyes were drawn to a photo pinned to the wall - an old black and white snapshot of the two of them, both looking much younger, arms around each other's shoulders, leant casually against a PM table, grinning widely for the camera. Ian took me through a door on the other side of the office that connected with the viewing suite, where relatives of those who had died in the hospital could request to see their loved one.

The suite consisted of an utterly bland but not uncomfortable little lounge area and the viewing room itself, decorated in a similar fashion to a funeral director's chapel of rest. It was in this room of course that the official identifications also took place; those scenes of human tragedy where the coroner's officer would lead the devastated relative into the room and ask them to confirm the shattering realisation that yes, the body was that of their son, daughter, mother, father, etc.

I was mentally recreating what it must be like for relatives entering the room to see their loved one when my thoughts were interrupted by Ian demonstrating how a separate hidden door allowed the body to be wheeled in from the body store area. He flicked a switch, demonstrating the two-tone spotlights: pink for routine hospital viewings and plain white light for identifications. At the time I was impressed with the logic behind the lighting arrangements, although later experience with that viewing suite would teach me that bereaved relatives lost out either way, because white light is a harsh and unforgiving way to display a dead body, whilst pink lighting can often make everything seem very lurid and theatrical. The limitations of the lighting arrangements aside however, Ian was at pains to explain that the reality of formal identifications bore no relation to the over-dramatic scenes depicted in films.

"The idea that the body is rolled out of a filing cabinet-style fridge," he explained dismissively, "Or worse still, that we flick back a sheet to show the body to the family, is pure fiction."

I nodded in agreement, but in fact I was secretly thinking the reality I'd witnessed that morning was infinitely more fascinating and grimly impressive than anything I'd ever seen in films. I helped Ian with serving a couple of funeral directors who had arrived to remove their newest clients, before Rick arrived as planned to collect two of our bodies and give me a lift back as well. But Ian wouldn't let us go without first telling us a joke about a lady and her husband's ashes, so before pushing the now empty body tray back into the fridge Ian used it as a prop for the kitchen table which featured in his joke. He mimed out all the actions on the newly-promoted fridge tray and delivered his punch line with great delight. Rick and I politely made a great show of laughing, but it was a rather lame joke (and certainly too crude to be repeated here).

All too quickly I was back at Thomas Broad & Son and back into the usual routine. Not many days later however, the morning car washing ritual was dramatically interrupted by fire. The oldest hearse and limousine on the fleet – a pair of Mk 6 Ford Granada conversions – were kept primarily for the hire work that the firm regularly carried out for a number of the builder/undertakers in the district. Despite not being old enough to drive at that time, I was still capable of, and permitted to, move the vehicles around the yard, so on that fateful day, having moved the limousine from the car washing area, I was about to roll out the matching hearse.

It was a cold morning and I thought the wisps of smoke coming from under the limousine's bonnet were probably just water evaporating off the warm engine. The wisps started getting thicker and became smoke. Not wanting to burst into the office only to find it was all a false alarm, I wandered in and casually said to Michael that there seemed to be "a rather worrying amount of smoke coming from the old Ford limo". He followed me back outside into the yard with a concerned look on his face. By the time we had got back out in the yard the engine compartment of the limousine was on fire, the heat having ignited a patch of paint on the bonnet whilst flames were also licking round the edges of the bonnet too.

The firm had its own petrol pumps at the top of the yard and Rick, who had also appeared, dashed up there to fetch one of the fire extinguishers kept near to the little hut housing the pumps. Meanwhile, the fire brigade were called as a precaution. Michael discharged the fire extinguisher and as the limousine disappeared in a cloud of white, fire-retardant powder, I could hear the sound of sirens down on the main road. For reasons I have never quite fathomed, a police car swung into the yard just moments before the fire engine. The premises were listed with the fire service as having petrol pumps on site, so the fire crew didn't take any chances and remained for at least half an hour afterwards, even though the fire was extinguished. Meanwhile the police officer in attendance tilted his head towards the radio clipped to the chest of his uniform and dourly announced that the incident was under control, that it was "just one of the black mariahs at the undertakers that had caught fire."

As we were all gathered round the stricken limousine the radio in the fire engine blared out an urgent message and at the summons of the engine's driver, the rest of the firemen quickly climbed back into the appliance and departed with the same flashing lights and sirens as they'd arrived. The policeman also took his leave, albeit less speedily. Michael, Rick, Paul and myself were left standing in a circle around the now completely white, extinguisher powder-coated limousine, parked forlornly in a large patch of equally white tarmac.

Paul looked at the car, then at me.

"A rather worrying amount of smoke... You silly little bugger!" he said exasperatedly, shaking his head as he marched back through to the office.

The sound of sirens was once more to herald a fire-bound incident in which the firm would become involved. Walking to work one morning, I was about to cross the main road at the foot of the funeral home drive when I was halted in mid-stride by the sudden appearance round the corner of a police car, blue lights blazing and siren wailing, hurtling towards me, quickly pursued by another patrol car. I let them whiz past and carried on my way. After getting into work and opening up the coffin workshop – my usual haunt at work – I could hear more sirens down on the road. There was a busy roundabout just past the funeral home, but judging by the scream of the engines above the sound of the sirens, I could tell that they had no intention of slowing down to approach the roundabout and that wherever they were going they were in one heck of a hurry.

By that time Rick had arrived and we settled into our morning routine. We had not long finished cleaning the cars designated for use that day before I headed down the back corridor towards the office to note down the details for the coffins that needed preparing that day. I had just put my hand out to pull open the swing door in the glass partition between the back corridor and the entrance hall when the door flew back at me, with Michael right behind it. I had barely regained my balance before Michael began barking instructions.

"Make sure the removal vehicle is kitted up and ready to go on a police removal and get Rick to bring it round and meet me out front. Someone's burnt to death in a road accident."

That explains all the sirens rushing past then I thought. Rick and Michael were gone for over three hours and even Rick, experienced as he was, came back looking rather pensive. As we set about the task of cleaning the heavily soiled removal equipment he told me what had happened. A car carrying two people on their way to work had collided with another vehicle and erupted in flames on impact. Whilst the passenger was dragged from the car, the driver was not so fortunate. Extricating the body was a complex task for the emergency crews and there was much grim speculation as to whether fate had granted the driver the mercy of being killed on impact. That would be a matter for the coroner's post-mortem to establish. The passenger also later succumbed to injuries sustained and we handled the funeral arrangements, whilst a funeral director to whom we regularly hired a limousine conducted the driver's funeral. Both victims were aged in their early twenties and both funerals had huge attendances.

Some years after the debacle with the limousine catching fire, I was to encounter another smoking limousine; this time during a period when, having gone self-employed, I was also carrying out limousine hire to other funeral firms. On one particular hire job on a blazing hot summer day, I was returning from Bath Crematorium and driving my passengers back up the very steep Weston Hill, out of the city. I noticed a few wisps of smoke and a burning smell in the cab and then the mourners in the rear slid open the glass partition screen and pointed out the same was happening in the passenger compartment.

Looking for a safe place to stop I glanced in my mirrors to check for traffic and was horrified to see white smoke billowing out behind my limousine. As luck would have it, other mourners were following in their own cars and had enough spare seats to collect up my now stranded passengers. Everyone took the incident in good temper (we were on the way home mercifully) although the mourner following me said he was glad I had stopped as he could no longer even see the road in all the smoke! The RAC were very efficient and helpful and the problem turned out to be a worn gasket leaking oil down onto the hot exhaust, from where it boiled off causing the smoke. The effects were far more dramatic than the problem itself....this time around!

One of those rite of passage moments that every undertaker has to face sooner or later is being involved with the death of someone they know. Often, the person may have been ill and the death expected. Such times are obviously sad, but at least you have the knowledge that you can make a very real and practical contribution. However, there are other times when, for us as undertakers, our personal and professional worlds collide head on.

I was barely months into my new job when, late one afternoon, my boss Paul and I were in the office alone and the telephone rang. I could tell by the way Paul spoke that this was someone he knew and that it sounded like they had suffered a tragic bereavement. When he finished the call he told me who the deceased was: a lad in his late teens who had been just a couple of years above me at school and had left at the same time as me. He had died in a road accident. Paul had just been talking to his father.

It was a deeply surreal moment when Rick brought this young chap back from the hospital and we lifted his broken body off the trolley. When I measured his body for a coffin I was surprised at how much larger he was than I remembered him. Engraving his name on the plate for the coffin lid was a strange sensation too: I was only sixteen at the time and I don't know why, but I suddenly felt very old. Here I was, carrying on living, whilst this lad would now forever be frozen in time, always just eighteen years old. Being in different years at school we obviously didn't have much to do with each other as a rule; but in that last year at school, being a senior pupil, he had once told me and my best friend off for being out of bounds skiving in a classroom when we should have been out on the games field. Now, less than twelve months later, here I was handling his dead body and furnishing his coffin.

As things turned out, he was to be the first of five lads (that I'm aware of) who were at school at the same time as me and who were to die tragically early. I would be involved with the funerals of another two. A year or so later a lad who had been in the year below me lost his life after drowning whilst on holiday abroad. Four years later, towards the end of my time with my employers, another lad from the year above me at school died in another tragic accident after falling from a building. By that time I was a qualified embalmer and after I had embalmed his body he was taken home to rest before the funeral. Again, another older pupil who I didn't know that well and yet there he was, lying on the slab, entrusted to my intimate physical care, in circumstances neither of us could ever have imagined just a few short years before.

With my recent experience of watching post-mortems now under my belt I felt well prepared when, one Tuesday lunchtime, Michael marched into the workshop and told me to stop what I was doing and get ready to go out with him on a police removal: a fatality on a railway line on the edge of town. I felt a sudden thrill of excitement and anticipation. I had seen one railway death after the body was brought into our mortuary prior to the funeral, but this would be the first time I had actually attended a railway scene itself. The death had occurred on a level crossing, although at that stage local officers from the British Transport Police were unsure whether the lady's death was accident or suicide. Our firm averaged a couple of such call-outs each year and time was always of the essence as, if there was no evidence of suspicious circumstances, there was pressure on the Transport Police to have the body recovered as quickly as possible and get the line re-opened.

When we arrived, a local GP who had been called to certify the death was still at the scene. From where our removal vehicle was parked in the lane leading to the railway line, we mounted the concrete steps up to the level crossing. It was drizzling with rain when we arrived on scene and I can always remember stepping up to the crossing, my mind full of imagined scenes. Not seeing anything on the level crossing, I instinctively looked up the line, only to see the doctor – whose presence was of course slightly unnecessary given the circumstances, picking his way over the shingle between the rails, one hand in his pocket and the other hand holding a little pink umbrella I can only assume he must have hastily borrowed from somewhere. As he approached Michael and I, he said quite cheerily,

"Hello chaps. I think they're ready for you. You don't need to be a doctor to tell that the poor woman is dead, but it's all about following the proper procedures, isn't it?"

Michael chuckled in agreement. Pointing to a cluster of police and rail workers further up the track the doctor continued,

"She's up there... Well, most of her anyway."

No sooner had we walked up the track with the stretcher than we were warned another train was on its way on the other line. We had to quickly cover the contorted body and then together with the police and rail workers, stand in a tight line in front of it, so as to further shield it from the view of the passing train's passengers who were already hanging curiously out of the windows as their train trundled past the scene at an enforced slow pace.

With their train past us we began the removal, making our way slowly back down the track and stopping once or twice to recover body parts. I think I'd imagined finding a scene of almost comic book grisliness, but just like my first post mortem, when the time actually came there were tiny details I could never have imagined, but which became so horribly obvious when I could see it all for real. The other thing that struck me at a more emotional level was how the body and its remains looked so utterly pathetic – in the true sense of the word.

We transferred the woman's body over to the public mortuary at Gloucestershire Royal Hospital for the subsequent post-mortem and within a week her remains were back with us again in readiness for the funeral. Michael dealt with the arrangements and mentioned to Rick and I that the family regarded the death as suicide, a view later confirmed when the eventual coroner's inquest recorded that same verdict.

There was only the one chapel of rest at Dartford House - Thomas Broad & Son's main funeral home and I was forever wheeling coffins in and out of the chapel for different families to visit. Often there were times when there would be one coffin in the chapel for, say, a 2.00pm viewing, another coffin in the adjacent bearers' room ready to be put in for the next viewing at say 3.00pm and finally the 4.00pm waiting on a trolley in the mortuary. On one such day I came wandering down the rear corridor towards the entrance hallway. As I pulled open the swing door Michael was stood there on the other side. He made a "ssshhh" gesture with his hand and whispered,

"Keep it quiet when you go into the office, the Smith family are in the chapel."

I looked at him for a second and whispered back,

"Don't you mean Brown? She's the one in there at the moment. I was told Brown now - at two o'clock - and then Smith at three o'clock."

Michael looked completely pole-axed. He bundled me back through the swing door.

"I've just shown the Smith people into the chapel" he hissed.

"But Brown is in there at the moment, like I was asked to do." I responded in a flustered whisper, so fixated with what I had been told to do earlier that the ramifications of what was unfolding now had yet to dawn on me.

"F***ing hell. Now what am I going to do?!" demanded Michael.

"I don't know." I replied, stung by the injustice of being blamed for doing nothing more than following instructions. I never could understand why such a busy company as ours only had one chapel anyway and as far as I was concerned this just proved my point.

"I'll have to go in there and speak to them." Michael decided. Then he hesitated, remarking thoughtfully "Mind, they've been in there five minutes already and they're very quiet." He glanced back at the swing door. "It's strange they haven't come out and said anything. Sod it, I'm staying out here. I'll wait for them to come out." he concluded.

As the two of us stood there, hidden behind the door, inspiration suddenly struck me. In a whisper I suggested that if we went into the bearers' room then we could listen at the door that led directly into the chapel. We might be able to gauge what was going on in the chapel then. Pressing our ears against the connecting door from the bearers' room, all we could hear was the sound of the two elderly ladies simply chattering quietly away in the chapel, but we couldn't make out what they were saying. There were certainly no raised voices or any hints of distress. We heard the sound of them opening the chapel door. Michael stepped out of our hiding place and back through the connecting door into the entrance hall to meet them as they came out of the chapel, beckoning me to follow. I walked innocently past the three of them and waited around the corner in the passage leading into the office - out of sight but still in earshot.

"Thank you very much. You've made her look very nice. We'll see you on Friday at the crematorium." And with that the ladies were gone. I reappeared. Michael was just stood there with a look of complete bewilderment on his face. We never did find out if the two elderly ladies who visited genuinely didn't recognise that they had seen the wrong body, or whether they were just being very British about it and were too polite to say anything.

The two of us felt like guilty schoolboys and we never mentioned the incident to anyone else in the firm, although later in the day Michael gave me another dressing down for putting the wrong coffin in the chapel. Having had the rest of the afternoon to reflect on events I realised it was possible that I might have got my instructions from the office mixed up, but I didn't tell Michael that!

We live in a very mobile world nowadays and so a situation that our firm dealt with a number of times each year was that of a death occurring abroad. Usually it was people who had died whilst on holiday, but of course there were the odd business travellers and on one very tragic occasion a newlywed on honeymoon. The actual causes of death tended to be split equally between natural causes and accidents of one kind or another. In the vast majority of these cases, the repatriation procedures were automatically enacted through the deceased's travel insurance policy. There are a number of funeral companies, naturally mostly in the London area, with specialist repatriation departments and two of these companies in particular were contracted to many of the main tour operators and travel insurers. In practice this meant that the repatriation contractor would provide a complete service including not only organising the repatriation itself but also collecting the coffin from the airport and delivering it directly back to the appropriate funeral director.

The more popular holiday destination countries cropped up time and again in repatriations, such as France, Spain and Italy, but in addition over the years I have handled bodies coming back from the USA, Canada, Thailand, South Africa, The Canary Islands, Sweden, Israel, Saudi Arabia and Germany.

We would receive notification from the family that the death had occurred and shortly afterwards the repatriation company would inform us when the body was due back in this country. The process normally took about five days. When the coffin or casket eventually arrived it would be my job to open it up. There are many consular and airline regulations surrounding repatriations, chief amongst which are the requirements that the body has to be embalmed and that the coffin has to be hermetically sealed, i.e. zinc lined.

Over time I would be faced with all manner of foreign coffins and caskets, some of which would be re-used on the funeral itself where the family agreed to this, whilst at other times the coffin would end up as handy pieces of wood for whichever of the staff wanted it for some DIY job at home. But it was the zinc lining that would usually provide me with the most bother. We always had a stout pair of metal shears in the workshop for those occasions and I would usually have to open the zinc lining tin-opener style, always after firstly taking a sniff when the initial hole was large enough, to see if there was any hint as to what condition the body was in.

The trouble with foreign countries is that they tend to have widely divergent views as to what constitutes embalming. In some countries the arterial injection method is practiced - with bodies arriving from the USA of course, but also Europe, generally in very presentable condition. However, in some countries embalming can only be carried out by a doctor, as opposed to a trained embalmer and the quality of treatment can vary enormously, ranging from the use of cosmetically harsh methods intended to preserve bodies for anatomical dissection, through to nothing more than just a small injection of disinfectant into the trunk of the body.

I vividly remember one body coming back from Thailand. I had the first hint of what was to come the moment I pierced the seal of the zinc liner. The wider I opened the liner the worse the stench became. The smell was so bad that I could actually feel a burning sensation at the back of my throat. There had been no sensible attempt at embalming the body, just a superficial soaking in disinfectant, but this had no useful effect whatsoever and the body was so decomposed that it was literally starting to slosh around inside the metal liner... The mixture of chemicals and decomposition gases meant that even a mask was of little use and so we could do little more than simply hold our breath, lift the body into a body bag and cover it with a liberal application of embalming compound – a powdered form of preservative nearly as astringent to the nose as the body itself was. Then all I could do after that was simply ensure that I put extra plastic lining in the coffin and seal down the lid as soon as the objectionable body was safely inside.

Funeral firms based in very ethnically mixed communities will tend to handle many outward bound repatriations, although in our corner of Gloucestershire the only common country we exported bodies to was Ireland, whether North or South. The airline regulations are a little different for repatriations to Ireland, omitting the need for zinc-lined coffins, requiring instead only that the body be embalmed and the coffin be securely padded and wrapped in hessian. We would wrap the coffin in wadding to protect it before my boss Paul's wife, Elizabeth, sewed up the hessian around the coffin using stout thread. The coffin would then be delivered to the cargo warehouse at the relevant airport. There is something rather odd about seeing a coffin amongst all the other cargo stacked and waiting to go.

On a number of occasions past and present, we've had repatriations where, rather than having an inclusive service with a specialist repatriation company, the body has been prepared and put on a flight and we have then been given the flight details by the family and left to arrange the import side of things ourselves, including collection from the airport. Particularly at somewhere like Heathrow, the cargo village - like the rest of the airport – can be a fascinating place.

Whether delivering or collecting coffins, the documentation accompanying the coffin must firstly be cleared and then it's round to the warehouse itself. The coffin – always enclosed in some form of protective covering or crate, but still instantly recognizable as a coffin – is plucked from the shelves in the warehouse by a forklift truck and brought out to your vehicle. Even to an experienced undertaker like me the sight of a coffin being driven through a warehouse on a forklift truck is rather bizarre, but I suppose to the forklift drivers even a coffin is just another item of cargo.

CHAPTER SEVEN

A Healthy Shade Of Pink

"He would make a lovely corpse."
Martin Chuzzlewit (Charles Dickens)

Having reached the end of my first year on the Youth Training Scheme, it was decided that I would remain on the scheme for a second year, taking up the option of finding my own training placement outside of the workplace for one day a week. Using a bit of lateral thinking, I was able to join - effectively unofficially - with that year's intake at the school of embalming that I would later study and qualify with. This rather neat arrangement enabled me to sit in on classes at the school and gain valuable experience for when I would enrol as an official student on the embalming course a year later. Meanwhile, this arrangement meant I could leave the YTS behind in all but financial terms and get my teeth stuck into proper work–specific training.

The West Midlands School Of Embalming, based in central Birmingham, was run by a former West Midlands funeral director who went on to run an embalming supplies company. A very well respected member of the British Institute Of Embalmers and accredited tutor, he ran his school by using the training room and mortuary facilities at the central office of a large funeral chain serving Birmingham and the West Midlands.

Although my employers were one of the largest funeral firms in Gloucestershire, seeing the principal funeral home of a large corporate company was a real eye-opener for me. Located just a mile outside Birmingham city centre, the funeral home not only served its own local client base, but also acted as the nerve centre for a network of branches across the city and beyond. Unlike our company, where everyone had to multi-task, all the staff at this funeral home had their own set duties, whether as funeral arrangers or conductors, chauffeur/bearers, mortuary staff or coffin workshop operatives and the only common denominator throughout the building was the echo of broad Brummie accents.

Behind the reception area was a long corridor with a row of chapels of rest, whilst our mortuary would have fitted into theirs half a dozen times over too. In the garage, itself the size of a small aircraft hangar, rows of gleaming hearses and limousines were parked, waiting to be sent out each day to service the network of different branches, whilst a fleet of three removal ambulances shuttled in a constant stream of bodies. The guys in the coffin workshop had templates for a bewildering array of different coffin styles and armed with industrial grade power tools they churned out a constant supply of finished coffins, ready for the mortuary staff to fill. But, for all its impressive scale, what struck me most was that it all seemed so soulless and anonymous. The whole place had the feel of a well furnished industrial unit and it lacked the character of Thomas Broad & Son's slightly musty, but venerable funeral home, Dartford House.

This was a time when I was beginning to understand that the grass isn't always greener on the other side. Yes, I was fed up with endless car washing and coffin fitting, but in fact I was doing many other things besides and much as I was impressed with the size and the dynamics of this enormous funeral home, when I actually tried to imagine working solely in the mortuary or solely in the garage I realised that in fact I wouldn't have had the scope that I currently had. Likewise, working in a small firm, each of our funerals had a sense of identity and importance; we knew the needs and wishes of each of our client families and also which funerals required particular attention, either because the circumstances were especially tragic or because the funeral was to have a particularly high profile locally. That's not to say that my employers treated some funerals as more important than others, but during the course of my weekly visits to that huge Birmingham funeral home I would see no concessions made for individual circumstances.

I would watch with frustration as absolutely tiny babies' coffins were driven away in hearses as if they were just very small adult coffins and on one occasion the embalming class watched a practical demonstration on the body of a murder victim, simply because he just happened to be first on the list of bodies awaiting embalming that day and so he was sent through as the token weekly body for the embalming students. There was nothing inherently wrong with that, but I just knew that had it been Thomas Broad & Son hosting the embalming school they would never have allowed a high profile case like a murder victim to be seen by anyone other than appropriate staff within the firm.

On embalming school days – held every Wednesday, the morning session would consist of theory tuition in an upstairs training room. I listened in as the students, some from branches of the host company and the rest from funeral firms across the Midlands, Wales and the South, were taught the basics of human anatomy and physiology before delving into the mysteries of the chemistry of the embalming process, together with the effects of various conditions and fatal diseases that an embalmer might expect to encounter amongst his or her caseload.

The lunch break would see the tutor and his students convening in a small, street corner pub just a few yards from the funeral home, before the afternoon was spent in the funeral home's busy mortuary, watching a practical demonstration. The company would always have a plentiful supply of bodies awaiting embalming by their staff and thus each week one body would be reserved for the school to embalm instead. The mortuary had the luxury of a separate embalming room, with very modern facilities, like powerful air ventilation and an adjustable height table (in our mortuary, the old slab was at just the wrong height to give you back ache after an hour bent over a body, whilst a high ceiling and a strong stomach had to take the place of ventilation). It was in this well appointed embalming room, on my first day joining the class – that about a dozen student embalmers, together with me as observer, would gather to watch the tutor demonstrate the embalmer's art on the frail body of an elderly lady.

Up to that point I had only seen a couple of embalmings being carried out by my boss Paul, himself a qualified embalmer. A family had requested that the body be returned to the house for a few days before the funeral and I had watched as Paul had carried out the treatment. Embalming is basically a chemical treatment, its aim being to achieve both a natural life-like appearance whilst also ensuring temporary preservation of the body.

The treatment itself consists of injecting embalming fluid (a mixture of formaldehyde, and other accessory chemicals, coloured dyes and water) into the arterial system of the body. The embalming fluid, injected via an electric pump, flows into the arteries under low pressure and this helps the fluid reach all parts of the body and penetrate into the tissues. The fluid reacts with proteins in the body to "fix" and harden the tissues, halting bacterial action and retarding the natural process of decomposition. This in turn prevents the unpleasant appearances, leakages and odours that might otherwise result. Meanwhile, to relieve the pressure created in the circulation system from the injection, any blood remaining is drained from a vein accompanying the artery being used.

The chemical composition of the embalming fluid can be altered to combat various conditions such as certain cancers, some diabetic conditions or drugs used prior to death, any of which can cause more rapid and significant deterioration of the body. Accessory chemicals can also be injected by hypodermic syringe directly into other areas of the body to supplement the overall treatment.

The arterial injection treats the main body structure: skin, muscles, and the organs themselves. However, after the main arterial treatment, the contents of the internal organs must also be treated, to prevent bacteria and gases building up and spreading to other parts of the body, causing decomposition problems even after arterial embalming. To prevent this happening, "cavity treatment" is carried out. A long, hollow tube, called a trocar, is introduced into the abdominal and thoracic (chest) cavities via a tiny incision just above the navel. Again using the electric pump, liquid and gaseous waste is aspirated (suctioned off). Then, again through the trocar, preservative chemicals are introduced using simple gravity pressure. Once all the chemical treatment is completed, the body is washed, dressed, shaved if it's a male and the hair combed or styled.

In cases where there may be superficial but unsightly discolouration such as bruising or abrasions, light cosmetic restoration can then be carried out with the subtle application of make-up designed for mortuary use. Where there has been significant damage or trauma to the facial area, a more complex restorative procedure will be carried out, sometimes taking many hours and involving the use of derma-surgery techniques as well as materials such as cosmetic wax and Plaster of Paris, before a cosmetic application to restore a natural appearance.

Thomas Broad & Son did not embalm bodies on a regular basis, using the treatment only in cases where the body was to be repatriated abroad, going back to the house before the funeral or if the funeral was to be delayed. Some of the minor techniques the embalming tutor demonstrated, such as the invisible suturing to hold a mouth closed so that a natural appearance was achieved, were revelations to me and a vast improvement on the techniques I had so far been shown by Paul and Michael. I would practice these techniques whilst dressing bodies in our own mortuary, with vastly improved results.

Our company also carried out routine embalmings for other funeral directors when required, but in addition to funeral firms we also carried out embalmings for a local assembly of the Plymouth Brethren, based in Gloucester. The Brethren, a Christian sect, always performed their own funerals. One of their representatives, who handled all their funeral arrangements, would bring the body to us in an old removal shell he had bought off my employers years before. We would then embalm the body and place it in a new, plain coffin. The Brethren representative would collect the coffined body again, together with the now empty removal shell, take the coffin back to the Brethren's meeting room on the outskirts of Gloucester and a simple funeral service would subsequently take place.

Other funeral directors would sometimes encounter something that we were better equipped to deal with and ask us to assist. During my second year, whilst I was sitting in on the embalming school classes Paul had told me that a funeral director from Gloucester was bringing over a body that had swollen to such an extent that he couldn't close the coffin lid. I was absolutely intrigued by this, knowing instantly that watching how Paul dealt with this body would add to my gradually enlarging fund of knowledge. I had heard the embalming tutor tell the students about severe swelling cases and I was pleased that I would get a chance to see the real thing.

Neville Small, Funeral Director, was based in a small village located on the west bank of the River Severn, on the opposite side to Gloucester. The village, which stretched out along the main road that ran from Gloucester to Chepstow, was home to a mix of agricultural families and well-off commuters, together with an established gypsy contingent. For all that his immediate area was sparsely populated and largely agricultural, Neville still had a thriving business and was well known not only in the rural areas surrounding his village but also over the river in the city itself.

Paul had told me about Neville's strapping great son Cliff and how when they arrived Cliff would probably be able to carry the body in by himself! I didn't think any more of it until they delivered the body later that morning. An ageing Ford Transit van, tilting lopsidedly, swung into the yard. Neville Small got out of the passenger side and greeted Paul. Neville had wild flyaway hair and the broadest Gloucestershire accent I had ever heard. I noticed the van heave slightly and creak ominously as the driver got out and shut his door. With that the van suddenly balanced itself upright. An enormous young man ambled round into sight from behind the van. He was at least 6' 5" tall and built like a brick out-house. Paul obviously wasn't kidding when he said that Cliff was capable of carrying the coffin by himself.

The two of them pulled the coffin out of the back of the van and wheeled it into our mortuary. Once the lady's body was placed on the slab I could see the scale of the problem. Her abdomen was swollen to a huge extent. I had never seen anything like it and I genuinely thought she might burst at any moment. Paul showed me how to tell if the swelling was gas or fluid by gently pressing and moving the swelling with his hand. He said if there was a slight wobble or a "wave" effect it was fluid, but if it felt hard and unmoving it was gas. It turned out to be fluid. Paul explained that the lady had probably suffered from Ascites, a condition most commonly caused by cirrhosis and severe liver disease, causing a massive build-up of fluid in the abdominal cavity. It was agreed that Neville Small would collect the body the following day and I spent the rest of the morning eyeing the body warily, still genuinely frightened that it might explode.

Later in the day Paul came back out to the mortuary, retrieved a trocar and hand pump from the embalming kit and using plastic tubing connected the pump and trocar to an 80oz glass jar. I watched carefully as he took a scalpel and placed the point of the blade on the swollen abdomen just above the belly button. I held my breath as, with practiced ease, Paul pushed the scalpel downwards for the length of its blade. There was nothing more than a dribble of yellow fluid from the newly made hole and to my immense relief, no explosion. I started breathing again. He then gently pushed the trocar deep into the hole and began working the hand pump; the jar started to fill with yellow liquid and Paul said,

"See how the swelling is reducing already?"

He filled the jar and asked me to empty it in the sluice before started the process again. By the time he had finished he had filled the jar a total of three times – 240 ounces of fluid.

The woman's body was absolutely transformed and once we had dressed her in the clothing that Neville had left for her, she looked perfectly normal for her size and build. Neville Small was equally amazed and delighted with the transformation. He said the woman's children were very distressed by their mother's appearance at the end of her life and asked him if he could do anything about it. He knew they would be absolutely delighted that their mother had been restored to a normal appearance one last time.

By the next time that Neville Small brought a body to us for attention a year later, I was far more experienced and confident in mortuary matters and was also by then a properly signed up student on the embalming course. The body in question this time was that of a gypsy man who had lost his life in a road accident. Neville wanted me to cosmetically restore some very unsightly, but largely superficial, injuries to his face. The arrival of this body brought with it another unexpected new experience for me – my first contact with an American-style casket. At that time the UK funeral profession was noticing a rise in the popularity of American caskets. These enormous and ornate, luxurious-looking, caskets found particular favour with those communities who traditionally chose very elaborate funerals – West Indian families for example, but also gypsies and travellers.

Neville travelled over in his hearse this time, with the enormous casket only just fitting into the vehicle. I was surprised at how little the casket itself actually weighed and indeed, when I had a proper look at it later on I was surprised at how fake the whole thing really was. The ornate metal casing was just that; a thin, hollow shell, with a sprung mattress and adjustable pillow inside. The "luxurious" velvet lining turned out to be moulded plastic panels with a suedette finish. The other thing that struck me was how feminine that particular style of casket looked, considering that its new occupant was, like all gypsy/traveller men, very definitely masculine. I carried out the cosmetic work and the huge casket soon left our mortuary again. It would be another few years before I would experience a gypsy funeral itself and what an education that would turn out to be!

Having now enrolled on the embalming course as a fully paid-up student, I spent nine months studying the theoretical and practical subjects, travelling up to Birmingham again once a week, before sitting a theory exam in the autumn. Having passed, I was then eligible to enter for the practical examination. Students were required to have carried out a minimum number of embalmings, consisting of 20 "straight cases" (bodies on which there had been no post mortem) and 10 "PM cases".

PM cases had always been a challenge for me, as the treatment of such cases was radically different to that employed for a straight case. During post mortem all the internal organs (viscera) are removed and inspected by the pathologist before being placed back inside the body. And therein lies the problem, because it means that the whole circulation system is destroyed. To embalm a PM case the viscera has to be removed again and separately treated with cavity chemical. The severed ends of the arteries serving the head and neck and the four limbs then have to be located in the empty trunk cavity and injected individually, but actually locating anything you could inject was often a challenge due to minute damage caused during the post mortem. However, in preparing for the practical exam my salvation was to appear in a most unlikely form.

My colleague Rick had often told me grim tales of his former boss, Mr. Summers - a very fierce and intimidating man by all accounts. However Rick would temper his stories by going on to say that for all he regarded Mr. Summers as a tyrant, he nevertheless respected him for setting high standards as a funeral director. Mr. Summers was also a keen advocate of embalming according to Rick and he would personally carry out all the treatments himself each morning before the office opened. At Mr. Summers' behest, a well-equipped mortuary was installed at his funeral home, complete with an expensive, stainless steel table of the type used in post-mortem rooms – a shiny altar at which Mr. Summers could worship the science of embalming, as Rick often remarked to me. Mr. Summers was approaching retirement, but just before that time came a stroke of luck would intervene for me.

Rick and I were sent to remove a body from the Gloucester funeral home where at that point Mr. Summers was still in harness as manager. He made sure he was there when Rick and I arrived, interested to see how his former staff member was getting on at Thomas Broad & Son. Knowing I would be interested to have a look around, Rick asked Mr. Summers to give me a guided tour of the funeral home and as we wandered round the building the conversation turned to my ambitions.

Mr. Summers was particularly interested in my progress as a student embalmer and to my utter amazement he offered to give me some unofficial tuition. To achieve the required number of cases that would enable me to do my practical exam, I had permission from Paul to embalm bodies that were not scheduled for viewing by relatives. So, for two consecutive Saturday afternoons Mr. Summers very generously came over to our mortuary to give me master classes on both a straight and a PM case. I'm sure any normal teenager would've had a million and one things they would rather do on a Saturday afternoon, but me being me, I was perfectly happy to be spending the afternoon in our mortuary, working under the experienced tutoring of Mr. Summers.

I believe he came from Swansea originally and he certainly had the broad Welsh accent to prove it, to the extent that he constantly referred to "embaaamin'." After all the stories Rick had told me about how strict and unforgiving he was as a boss, it was with some trepidation that I waited in the funeral home yard for him to arrive. I needn't have worried. Mr. Summers proved to be an excellent tutor; patient, friendly and with a very down to earth way of explaining things. He taught me a huge amount, and there were more than a few moments under his temporary tutoring where "the penny dropped" for me on certain techniques.

I particularly remember the second of our Saturday afternoon master classes, when he guided me through the challenge of embalming a PM case, giving me wise advice about how to approach the locating of arteries in those cases. I had prepared the body half an hour before Mr. Summers arrived. I had taken off the skull cap – originally removed during the post-mortem and I'd also removed the internal organs once again and left them to soak in embalming chemical.

Now, with the two of us leaning on the rim of the slab, contemplating the opened-up body laying in front of us, Mr. Summers' broad Welsh accent echoed round that large, white-tiled room as, without the least trace of irony, he said:

"Now Jaaaymes, you 'aven't to be frightened of a post-mortem case. You need to remember that you caaan't do any more damage than the pathologist 'ave already done."

There are six principal arteries that embalmers normally use when injecting fluid, but in cases where the fluid circulation to specific areas of the body is problematic, the embalmer can then use a range of secondary injection sites, with the radial artery in the wrist being one of them.

A demonstration in class from my embalming tutor in how to raise the radial artery was later to prove really helpful when dealing with an instruction that one deceased lady had left for her family. The fear of being buried alive is an old one and in not so distant days people would often ask their doctor to sever an artery after they had died, just to ensure that they were indeed dead. Sure enough, we were to receive just such a request from a family: to carry out their elderly mother's last request and sever one of her arteries. It was at least a chance for me to practice my radial artery technique and having made a very small incision in the wrist and raised this tiny artery, I tied it off with surgical thread at the two ends of its exposed length and then severed it in between. Since then I've had two or three more identical requests and carried out the same procedure again.

An altogether less fiddly procedure is the removal of pacemakers. These devices, no more than 5cm wide and less than a centimetre thick, are nevertheless capable of causing a damaging explosion when exposed to the kind of heat a cremator furnace produces Most modern pacemakers have a lithium/iodine-polyvinylpyridine (PVP) battery—because of its greater longevity and smaller cell size, but other power sources have included zinc/mercuric oxide, nickel cadmium and even plutonium-238! The first reported case of a pacemaker explosion during cremation was apparently in 1976. The body of a 70-year-old man was cremated at 800 °C. After 5 minutes, four small explosions occurred in rapid succession with a final explosion a few minutes later. Although the damage was confined to the fire bricks within the furnace, later cases have resulted in greater damage to equipment and even injury to crematorium staff.

Because of this danger, a statutory question on the medical cremation certificate which the deceased's doctor signs, asks whether the deceased has a pacemaker and if so whether the device has been removed. In the case of hospital deaths, or when a post mortem is carried out, the pacemaker is automatically removed before we take custody of the body. However, for deaths occurring in private residences or care homes, then the responsibility for ensuring the pacemaker's removal rests with the deceased's GP and ultimately the funeral director as well.

Our local GP's usually rely on us to actually remove the pacemaker, which involves nothing more than a small incision in the upper left breast area, where the pacemaker will have been superficially located underneath the skin. The pacemaker can then be easily removed. Pacemakers are not the kind of thing you can put out in the rubbish and instead we have an arrangement to return them to the hospital mortuary from where they're passed back to the Cardiology Department.

My practical embalming examination came a few months after my unofficial lessons with Mr. Summers and the exam was hosted by another large funeral home, this time in Bristol. With a racing heartbeat, but with feet of clay, I carried my kit into the embalming room. Remembering my embalming tutor's horror stories about over-zealous examiners I had brought a manually operated fluid pump, in case the examiner saw me with an electric one, switched it off mid-treatment and announced "You've had a power cut. Please switch to manual operation and carry on." However, in the event my examiner was a very friendly, easy going chap and he immediately put me at my ease. My joy increased when, pulling the hospital shroud off the body, I saw that I'd been given a straight case to embalm.

Eventually I got the letter in the post to say that I had passed the practical examination and that consequently I was now a qualified embalmer. In professional circles I was now entitled to use the letters "MBIE" (Member of The British Institute Of Embalmers) after my name.

Witnessing Paul's trick with the trocar on the swollen body a year or so earlier was to bear unexpected fruit for me. Myself and Stan - one of the principal part-time bearers, had done a removal for the Dursley branch office and our instructions were to deposit the body there for the GP to go in and sign cremation paperwork and also to bring back another body already in storage that the branch receptionist had said was starting to "go off." The receptionist had clearly understated matters, as we found a very swollen body waiting for us. Looking at the body, Stan's face registered the same shocked fascination as mine had when I had seen that first fluid-filled body. I felt the abdominal area: hard, no wobbling. This time it was definitely tissue gas, the result of the first stages of decomposition within the organs. I said to Stan that I knew exactly how to deal with that little problem and when he asked how, my explanation left him eager to see the procedure carried out when we got back to the mortuary at head office.

We had the car windows open all the way back and the body went straight onto the slab when we arrived. With Stan transfixed by what I was doing I pierced the abdomen with a scalpel and sure enough there was a gentle hissing noise as the pressure was instantly relieved. After that initial release I used the trocar to ensure complete removal of the gases and treated the affected area with embalming fluid to retard the worst effects of the decomposition. Stan was deeply impressed by all of this, whilst I – basking in glory of course – was just very relieved that I had made the right call.

There is only one circumstance where it is permissible to place two bodies in one coffin for either burial or cremation - when the deceased are a mother and her infant child. I have only once encountered such a situation so far in my career. William Stevens, one of the funeral directors to whom we regularly hired vehicles, brought to us a mother and baby who had died after a road accident.

William had provided an elegant English-style casket (basically, just a rectangular coffin – far less ostentatious than its American counterpart) and all that was required of me was, as usual, to dress the bodies, arrange them in the casket and then deal with the cosmetic restoration. The accepted rule with facial injuries is, where possible, to tilt the face to accentuate the best side and with this thought in mind I examined the tragic pair laying before me. Through our company I was starting to gain a reputation for being able to deal with awkward cases and it was because of this reputation that I was being given the chance to become involved with things that I might not otherwise have encountered.

The down-side was that Paul and Michael came to *expect* me to be able to deal effectively with those situations – situations that neither of them had ever had the time or, more importantly the inclination, to learn how to deal with themselves – and so there was pressure on me to deliver, when in fact I was breaking new ground for myself as well. So, with Michael and William both watching me, I examined the bodies. Fortunately for me, the pattern of injuries immediately suggested a plan and I confidently announced that I could do something, my words carrying a strong hint that I could be left alone to get on with it. The hint was at first lost on Michael, but William was expecting to be able to leave the bodies and come back later anyway.

Eventually I was left in peace to get on with things and I began by dressing the two bodies. My plan was quite simple: once the mother's body had been placed in the casket I would place her infant child in her arms, using generous amounts of hidden padding to support the positioning. The right side of the mother's face carried the least injury, whilst for the infant it was the left side, so I wanted to rest the infant's right cheek against the mother's breast, before tilting the mother's left cheek down against her child's head. This would create a very natural and maternal pose and along with further subtle cosmetic treatment, would also enable me to hide the worst of the injuries to both faces.

Having completed the embalming course and qualified, I was very keen to ride the crest of the educational wave and get straight on with studying for the Diploma in Funeral Directing. The course was based on the National Association Of Funeral Directors' wonderfully named "Manual of Funeral Directing" and the syllabus, like the manual, reflected best practice in funeral service.

Before the Diploma was awarded, students were expected to have gained a comprehensive knowledge of all aspects of the role of a funeral director and to have spent a minimum of two years within the profession. Candidates needed to have practical experience of conducting funerals and in addition needed to have personally arranged at least twenty five funerals. The final examinations themselves were based on a theory element comprising of four modules, eight reports and a written paper, whilst the practical part of the syllabus involved arranging a funeral, where the student was put in a "role play" situation, with the examiner acting as the bereaved client.

I studied for the Diploma at The South Western School Of Funeral Sciences in Salisbury, in my native Wiltshire, finally qualifying in 1991. For six months I studied for the course by distance learning, with a monthly educational session at the school itself. I used to look forward to those study weekends as the drive down there took me through the parts of Wiltshire that I remembered from my childhood: the fairly workaday town of Chippenham, where way back in 1977 my parents endured two consecutive attempts at queuing outside the Astoria cinema so that I could see "Star Wars" for the first time, before passing through the altogether more picturesque town of Devizes, where even now the town brewery still uses Shire horses to pull the dray carts, then heading through the cosy, "olde worlde" villages of Potterne and West Lavington.

From there I would travel out across the vast openness of Salisbury plain, before the first of the city lights could be seen twinkling in the distance. The plain is used as a military training range and dotted along the main road are tank crossing points, the road signs for which used to fascinate me as a kid – there aren't many places where you will see those red triangular road signs with a little black tank in the middle. Sadly, in neither childhood or adulthood did I ever get to see a tank.

The South Western School Of Funeral Sciences was attached to a funeral business, the directors of which were registered tutors for both the embalming and diploma courses. Their company office occupied a large Victorian house in a residential street on the edge of the city centre, whilst their garage, coffin workshop, mortuary and chapel of rest were housed in separate premises in a neighbouring road. My monthly tutorial sessions would be spent in a portacabin classroom in the rear yard of this building.

The course saw me embark on a learning route through a mass of different subjects, principal among which were the law of burial, cremation and exhumation, religious rites, special subjects such as repatriation and burial at sea as well as the process of actually arranging and conducting funerals, health and safety, funeral administration and even wills and probate. After the initial induction weekend, when there were about a dozen of us, it transpired that only two of us were actually studying via the correspondence course. Everyone else would study with weekly tutorial sessions. That left me and another young chap from a firm in Bognor Regis as the correspondence students. That was actually quite nice, as with just the two of us and our tutor we were able to take our time at our monthly tutorial at the school really picking our way through the syllabus and sharing experiences and knowledge.

Away from the course my learning had been, to some extent, self-taught. Simply working in a busy firm and being immersed in a funeral directing environment had taught me a great deal and every time I was in the office collecting details for the coffins that needed furnishing, or checking which vehicles were on which funerals, I would be constantly listening and paying attention to everything around me. Just by listening to Paul, Michael & Muriel the secretary making telephone calls was an education in itself, taking note of how they spoke to families, vicars, doctors, coroner's officers, the terminology they used and the way they expressed different things. Likewise handling paperwork, either to find the information I needed regarding coffin requirements or jewellery instructions, or when delivering and collecting official documents, all added to my education.

Needless to say I can vividly recall my first arrangement meeting with a family. It was a house visit and the family were the kind of people who make our work a pleasure: easy going but with a clear idea of what they wanted to achieve and thankfully armed with some clear requests left by the deceased too. I will never know if they realised how nervous I was, but I managed to remember everything I needed to ask.

Looking back, it was of course all rather stilted and unpolished, lacking a more conversational approach to the meeting, but back then the only measure I had was what the firm expected of me and as long as I completed everything to the satisfaction of those in the office, then I had to regard the task as correctly performed. Paul and Michael both had their individual ways of talking to families, Paul tending to be very formal and professional, whereas Michael had an equally effective, but slightly more informal way about him. Of course practice and experience would prove to be the only route by which I would gain confidence in working with families during arrangement meetings and over time I would develop my own style.

The funeral itself was held in the chapel at Gloucester Crematorium – what we would term in the office as a "straight to the crem" funeral. As we approached up the drive I was pleased to see the familiar figure of Patrick the chapel attendant stood by the chapel door and less than a minute later there I stood too, in a borrowed tailcoat and striped trousers, trying to look calm and in control. I was mentally rehearsing all the little things that I had observed Paul and Michael do on funerals so many times before: gather up the mourners, establish the extent of the family group, lead them clearly to their seats and ensure everyone was seated and had a copy of the service sheet. For all that Michael in particular could get on my nerves at times, looking back on things now I can see that when it came to conducting funerals he had a remarkable degree of emotional intelligence and he was really good at handling people. He knew how to keep everything under control and marshal large crowds of mourners, yet still balance the practical requirements with the lightness of touch and sense of occasion that all good funeral directors have.

The entry into the chapel went smoothly and I was able to relax for twenty minutes or so whilst the service got under way. At the end of the service there was that equally awkward moment when having led all the mourners back out again, I had to give them time to mingle before encouraging them to move, hopefully as a group, across to the Garden Of Remembrance where the flowers were laid out for them to see. This was in the days before the crematorium had a separate exit from the chapel and it was a constant challenge for funeral directors to marshal their mourners in such a way as not to make them feel like sheep, but all the while keeping an eye on the main drive, watching the cortege for the next funeral drawing ever closer. That was one pressure we didn't have with burials, as almost invariably you have cemeteries and churchyards to yourself for the day.

There was an interesting footnote to that first funeral – interesting to me anyway. I dealt with the deceased's wife and daughter when I made the arrangements, but I remembered being introduced to the son-in-law at the crematorium. At the time a well known national utility company was in the news headlines and a television news report featured the chairman, who turned out to be the deceased's son-in-law.

I have already mentioned the requirement that before qualifying, Diploma course students had to arrange and conduct a minimum of twenty five funerals, or *frooneralss*, as Gloucestershire people are so beloved of saying. (I had some very rural relatives by marriage who often spoke of "Granny's *frooneral* having to be delayed because she was taken away to have a *post mortle*." Like I said, we were only related by marriage...).

Once again, Mr. Summers, who by that time had retired, was to be an unlikely ally. We had a chance meeting in the crematorium office where he was calling for old times' sake and this time around we got chatting about my Diploma studies. The result of the conversation was that, about a week later I found myself sat in Mr. Summers' kitchen whilst we acted out a role play funeral arrangement meeting. Once again he was able to pass on some of the fruits of his experience and help me to polish my technique.

The Diploma theory examinations took place in a conference centre in the West Midlands. A good proportion of the candidates had clearly been granted a few hours' time off work to come in and do the exam as the reception foyer was a sea of black jackets and striped trousers, whilst I could see more than few removal vehicles dotted around the car park. I recognised a young chap from a Gloucester company who I had often seen conducting funerals at the crematorium. So, by way of simply making conversation to fill those tense minutes before the exam, I said to him,

"I take it this is the right place for the funeral Diploma exams?"

"Well, either that or it's a Freemasons' convention." Came the rather tart reply.

I decided it was a good moment to keep myself to myself and I wandered off again.

The practical examination, a role play funeral arrangement with an examiner playing the part of the bereaved client, came later. I remember myself and about four other candidates all lined up in a hotel corridor, each of us stood facing the door to a room in which an examiner was waiting. We had to wait for a signal, then all knock on our respective doors and allow the "client" to answer the door. It reminded me of an executioner's memoirs I had read, where the executioner, the prison governor and the warders all stood on the landing of the prison wing, outside the condemned prisoner's cell, waiting for the signal to enter on the stroke of 9.00am – the usual hanging hour. From my point of view the situation wasn't so very different.

Eventually I had the notification that I had passed both examinations and finally I was able to tick off another ambition: to become a fully qualified funeral director and embalmer. I was now:

James Baker MBIE, Dip. FD, MBIFD,
(Member of the British Institute Of Embalmers)
(Holder of the Diploma In Funeral Directing)
(Member of the British Institute Of Funeral Directors)

It was 1991. My basic training was over. Now the learning would start.

CHAPTER EIGHT

Old Man River

"Ol' Man River, that Ol' Man River
He must know somepin', but he don't say nothin'
He just keeps rollin', he keeps on rollin' along."
(Lyrics by Oscar Hammerstein II, from the musical "Show Boat")

Tucked in the five valleys, Stroud sits right between the Cotswold Hills and the floodplains of the Severn Vale, through which flows the mighty River Severn, Britain's longest river. From its source in the Welsh Cambrian mountains, the river meanders in and out of the English counties of Shropshire and Worcestershire and Gloucestershire and along its banks sit the historic cities of Shrewsbury, Worcester and Gloucester. The river finally empties into the Bristol Channel, where the estuary forms a physical boundary between England and Wales.

The river is less than a 100 yards across at Gloucester, widening to a few hundred yards where it passes the western reaches of the Stroud district and by the time it reaches the southern edge of the county it is one mile across. Where the river empties into the sea just past Bristol it has created a massive estuary over five miles wide. The tidal range is the second highest in the world, being as much as fifty feet (approx. fifteen metres).

Meanwhile, in a unique reverse effect, these huge sea tides, flowing *into* this wide estuary and *up* the river, give rise to the famous Severn Bore, which despite its rather unpromising name is one of Britain's most spectacular natural phenomena. As the width of the river decreases rapidly the further back upstream you travel, so the depth decreases rapidly too, thereby forming a funnel shape. As the incoming tide travels up the estuary, it's routed into an ever decreasing channel and consequently a surge wave or bore is formed. This wave has an average speed of just under ten miles per hour and can reach two metres in height – the bore being graded by stars, with the ironically named "five star bore" being the most spectacular.

It was just such a five star wave that my school geography class was taken to watch one Autumn day in the early 1980's. I'd always regarded geography lessons as a five star bore anyway, so my expectations were not that high. But I was proved very wrong because, gathered on a riverbank viewing spot just south of Gloucester, our vantage point offered what turned out to be a spectacular view of the immense wave as it passed by.

The next time I would find myself clambering around on the riverbank a few years later it would be for something else that the tidal surges had brought up the river into my part of Gloucestershire.

Myself and a colleague were sent out one morning on a coroner's removal from the river, at Epney, one of the cluster of picturesque Severnside villages on the river's eastern banks, at the outermost edge of the Stroud district. I'd experienced many coroner's removals by this time, but this was my first river removal. A staff member from the nearby riverside pub had spotted what they thought might be a body floating out in the water and they had raised the alarm. The road that ran alongside the river bank was soon congested with police cars, together with the mobile control unit for the Underwater Search Unit from the neighbouring Avon & Somerset Constabulary.

By the time we had arrived at the scene the body had been brought to the shore. At this location the river is two hundred yards wide and for a virtual non-swimmer like me that great expanse of brown, murky water looked more threatening than picturesque. I suppose if you don't mind water then drowning yourself is probably no worse than say, hanging yourself, but stood on the bank looking at the water I could not imagine a worse fate than drowning. I watched the police divers with great admiration. This looked like a thoroughly dangerous place to be messing about on the water.

The body was that of a middle-aged male and clearly hadn't been in the water for more than a few days. That isn't to say however, that whichever funeral director ended up arranging his funeral would not have had a trial on his hands. A body immersed in river water for even the shortest time will begin to decompose very quickly, because of the bacteria in the water and by the time coroner's formalities had been attended to, that body would have been thoroughly offensive to all who handled it.

Although dirty and disheveled now, I could see that the dead man had originally been dressed very smartly, as if he'd been out at a black tie social function. In fact, according to the police officers present, the unfortunate victim had fallen into the water during a party at a river-side location on the western side much further downstream near the Welsh border. The strong upward tide, moving against the prevailing downward flow of the river, had carried his body all the way back up to the waters by Epney where it was finally spotted and recovered. It was a tragic, accidental drowning but, given his posthumous journey, it was a mercy that his body was delivered up again by the river tides, allowing at least the chance for decent burial.

My next trip Severnside was to Longney, the next village up the river. Late one autumn Saturday afternoon, my boss Paul 'phoned me with instructions for another riverside coroner's removal. He gave me the location and could tell me only that the police wanted us to remove an unidentified body. He went on to say that he would tell Stan, the part-time bearer who was on call with me that day, to meet me down at the funeral home. I was actually quite pleased – after all I was on call anyway and this would be far more exciting than just a routine trip to a nursing home. I only had a short walk down to the funeral home and whilst waiting for Stan I made doubly sure the removal vehicle was well stocked with disposable body bags, gloves, aprons and two pairs of wellie boots.

In stark contrast to my last visit to the river there were no great scenes of activity when Stan and I arrived. No crowd of officers and patrol cars, no underwater search unit. In fact there was just a solitary police car parked on a wide verge next to the path that led down to the river. *They've obviously ruled out any possibility of suspicious circumstances then,* I thought to myself. As we parked up a solitary police constable appeared, and a thoroughly disgruntled-looking police constable he was too. He had the air of someone who intended to be of as little help as possible.

After a curt greeting from the officer we all descended the path far enough to be able to see down onto the muddy shore below.

"It's over there." He said, without ceremony.

Following the line of his pointing finger to where "it" lay, I could see a mounded object just a few metres back from the water's edge. Apparently a curious dog walker, sensing that it didn't look like driftwood or other debris, had decided the odd-shaped object warranted a closer look and a 'phone call to the police quickly followed. I could readily imagine the dog walker's shock at the discovery. I've often thought there would be countless dead bodies lying undiscovered all over the country if it wasn't for the almost obligatory dog walkers who regularly make these grim discoveries.

"You'll need to be fairly quick about it, the tide will be coming in soon."

The officer's words jolted me into rapid thought as my focus switched from the body to the river itself. For a second time I found myself contemplating that threatening expanse of water, lapping quietly and innocently onto the muddy shore. I remembered watching the power of that bore wave years ago and it felt as though, in return for brightening up a dull morning at school, the river was now calling in the favour, anxious for me to relieve it of the body it had been holding in its murky depths. I didn't hesitate a second longer. I quickly summed up the situation, decided on a plan of action for the removal and turned back towards the car.

We didn't have disposable overalls back then, but I could tell that the disposable plastic aprons we carried wouldn't be of any real use either, so after Stan and I had put on wellie boots and gloves, we pulled out the stretcher trolley and folded it down onto the ground behind the car, ready to receive the body when we carried it back up on the flexible stretcher. In planning the removal my suspicions were proved correct in one vital respect: the police officer had no intention of going anywhere near the body. The trolley wouldn't have been any use on the sand and mud anyway, but I had been hoping that the officer might at least help us carry the flexible stretcher back up to the car. So with folded body bags in my hand and with Stan carrying the rolled up flexible stretcher, we marched back down the path. Any thoughts I may have had about what we would encounter when we actually handled the body were now overridden by the more immediate concern that the tide was due in and we were effectively on our own.

We trudged across the mud and had our first close look at the body. It had been in the water for a very long time and was in a very advanced stage of decomposition. It is quite common for bodies in water to be discovered only after the gases formed within the trunk bring the body back to the surface - such cases historically being referred to by the objectionable term "floater." In this case we were well past the floating stage and it was simply the tide that had left the body behind, just like driftwood.

The only remaining clothing was a pair of very shrunk trousers still on the body, complete with leather belt and from that I deduced that the body was that of a male.

Stan unfolded the flexible stretcher whilst I opened out the first of the disposable body bags and laid it inside the stretcher. It took a few attempts to manhandle the body into the second body bag as the slimy remains defied all our attempts to get a firm grip, but once it was into a body bag and we had something to hold onto, we heaved the bagged-up body into the second bag on the stretcher and zipped everything up. The double-bagging would at least help to make the journey in the car bearable as well as making life a tiny bit easier for the staff at the public mortuary.

In any other circumstances that spot by the river might have been quite a pleasant place to be on an autumn evening – the sky had a dusky glow, the air was utterly still and the only sound was that of the water gently but inexorably lapping further up the sand. However, we weren't there for an afternoon stroll and so, with no small amount of relief on my part, we began moving away from the water's edge. My relief was short lived.

Despite the body being partially skeletonised, the combination of its weight and that of the stretcher meant it was still surprisingly heavy and I had barely taken one step before my boot sank firmly into the mud. I had to put my end of the stretcher down again just to gain the leverage to liberate my boot from the mud and for a moment I had dark visions of being stranded in the mud with the tide rising towards me. It took a lot of heaving, huffing, stopping and starting, but eventually Stan and I reached the path leading back up to the car. We were sweating, out of breath and plastered in mud, but the police officer's facial expression was still registering the same complete disinterest that it had when we first arrived. One last blast of effort took us to the car and the welcome relief of the waiting trolley.

It was dark by the time we finally drove away and with the windows wide open we made our way to Gloucester Mortuary as rapidly as we could. We had already phoned ahead for the hospital switchboard to call out the duty technician and it turned out to be Gerry on call that weekend. When we arrived and told him what we were bringing in, his reaction was pretty much the same as that of the Fatted Calf when it was given the news that the Prodigal Son had turned up.

To the best of my knowledge I don't think that body was ever identified, despite police investigations and careful searches through lists of missing persons. The tidal nature of the river Severn is such that, as I've already explained, the body could have come from anywhere and although, at some point, the river would usually gave up its dead, the nameless man's exit from this world remained unseen and unknown. It would be quite a few years before I would be back on the river bank again, this time in pursuit of a body whose identity was definitely known and his exit from this world was not only well documented but also reduced to a grisly comedy of errors.

It was a Wednesday afternoon, between Christmas and New Year, when we took the call from police control, asking us to attend at a riverside location at the southern end of the county, near the historic town of Berkeley, famous for its castle, its hunt and the Berkeley family themselves. (History records that a hundred years ago the Berkeley family could go out hunting from their castle in Gloucestershire all the way to what is now Berkeley Square in central London, without ever leaving their own land.).

Once again it was a rush to get down to the location to make the most of the available daylight. I was aware from local news reports that a vehicle belonging to a man missing from the north of the country after being wanted by police in connection with serious criminal allegations, had been found abandoned near to the river on the outskirts of Gloucester. When I saw the reports I thought to myself *give it a few days and one of the Gloucester boys will have to go and remove him from the river bank.* However, whilst my assumption that he would be found dead in the river proved correct, the strong tides and currents had prevailed and so it would instead be *me* removing him from the river bank.

Upon reaching the given location, which, as is so often the case with coroner's removals, was somewhere we didn't even know existed, we were directed by a police officer to drive round the perimeter of a farmyard and park by a large embankment. Keith, my father-in-law, who was working as a part-time bearer during that time, was with me. We could see more police officers standing up on the crest of the embankment and as we walked up to the top, the river - at this point a mile wide, stretched out in front of us. On the opposite bank I could see the street lights of Lydney, a town situated at the southern tip of the Forest Of Dean. It was thirty five miles away by road, but close enough at that point that I could have counted all those street lights.

We were stood on a flood defence and below us were large boulders on the river bank. Once again, with my line of sight directed by one of the police officers, I was able to make out the body tucked amongst the boulders. It all looked fairly accessible from where I was standing and seeing that I could easily drive our removal ambulance up onto the embankment I made my way back down to the vehicle. As always with removals of this type, I selected the older, less valuable of the two sets of equipment that the ambulance carried and once again armed with the flexible stretcher we went in pursuit of the body.

Admittedly, wading across a mud flat was much harder and messier, but clambering over boulders is not exactly the easiest of tasks either. We found it very difficult even to unfold the flexible stretcher as there was no space to lay it flat, let alone try to manhandle the body onto the stretcher, all the while trying to find a safe footing on the slippery rocks. The body had been in the water for about a week, which by coincidence is roughly how long it then took for the smell to dissipate from our ambulance after this eventful removal was completed. The deceased was a well-built chap and very heavy. This time around a couple of the police officers lent a hand to carry the stretcher back off the boulders (they carefully timed their intervention to avoid the actual recovery and handling of the body of course). I think the four of us slipped off the wet, slimy boulders into little crevasses underfoot at least once each, with the attendant risk of sprained ankles, but we managed to get back up the embankment without injury.

When we arrived at the mortuary at Gloucestershire Royal Hospital a few C.I.D. officers were waiting for our arrival. There wasn't any intention of performing a forensic post mortem at that stage, but given that the deceased was wanted for police questioning before he disappeared I wasn't surprised to see C.I.D. there. But it was at that point one of the officers realised that no-one had arranged for the body to have been certified dead at the scene by a doctor. Despite being medically pointless, the certification of death is still always a vital legal requirement, but for whatever reason the procedure hadn't been organised at the scene as it should have been.

By this time the original mortuary technicians Gerry and Ian had both left their employment at the mortuary and so Derek, the technician on duty that evening, phoned the hospital switchboard and asked for an A & E doctor to be paged. When the doctor phoned Derek back we could overhear some negotiating going on before he put the phone back down.

"There's no one free to come round from A & E – they're too busy. The duty consultant said could you take the body round to the ambulance bay and he'll come out to certify it from the back of the vehicle." Derek announced.

Unable to believe what we were being asked to do, we reluctantly loaded the body back into our ambulance and drove back out onto the main road and round to the entrance for A & E. Bearing in mind it was now a late evening during the Christmas holiday period, the ambulance bay for A & E was predictably busy. I thought the presence of an undertaker's private ambulance might be an affront to all that the A & E Department stood for, so I parked an extra few metres away from the front entrance. Then we just had to wait....for twenty long minutes as it turned out. I just wanted to turn invisible because we were attracting more than a few odd looks from paramedic crews.

An emergency consultant eventually came out and explained that as soon as he'd seen the body he would phone back through to the mortuary and confirm to the waiting C.I.D. officers that he had certified the death. I opened the tailgate, pulled back the stretcher cover and unzipped the body bag. The consultant instinctively reached for the stethoscope around his neck as he leant forward into the partially opened body bag, but then his head shot back quickly, his hand quickly moved away from his stethoscope and he screwed up his nose.

"Oh god yes. He can't be much more dead than that can he?! Ok, thanks for doing that. You can whizz him back round to the mortuary now."

Still stunned at having to go through such a ridiculously insensitive and undignified exercise, it was with great relief that we made a second and finally successful attempt at depositing the body back in the mortuary, leaving C.I.D. with their wanted man - now officially deceased. From the local news reports of his disappearance I believe he was wanted for questioning on suspicion of offences against children.

CHAPTER NINE

Remembrance Sunday

"Murder is unique in that it abolishes the party it injures, so that society has to take the place of the victim and on his behalf demand atonement or grant forgiveness; it is the one crime in which society has a direct interest."

W.H. Auden

By the Autumn of 1989 Rick had left the firm and returned to work for his previous employers in Gloucester, filling a vacancy that had arisen for a qualified Funeral Director. I was left effectively working on my own "round the back" for the next year or so, most notably during the dreadful January of 1990 when, due to a flu epidemic, our company conducted a mind-numbing 83 funerals in that one month alone. Most people would say "Gosh, I bet you guys were rubbing your hands with glee – all those bodies, all that money!" The snag is that most people forget that before the money rolls in, there is one heck of a lot of work to be done to earn it. The fridge was permanently up to its unofficial 12 body capacity, the slab was never unoccupied and I often had bodies lined up on the mortuary floor, as well as placed in coffins and parked in the bearers' room to await attention. A doctor who came to see a body in order to complete cremation certificates likened the scene out the back to "something like the fields of Flanders." There was absolutely no respite and as soon as one body left on a funeral there would be another two coming in to replace it.

We even reached the stage where our coffin suppliers, of which we had two or three, just couldn't keep up with the demand from all their customers and deliveries were being rationed, as the factories were already running at full production capacity. However the lunacy eventually passed and things returned to more manageable levels. But during the months between the Summer and Autumn of 1989, two rather significant events occurred for me.

The first event came during the late Summer of 1989, when another young work experience lad came to work for us. He was also a James and so was quickly christened within the firm as "t'other James." The similarities ended there, as t'other James was tall, dark and sporty, whereas I.... wasn't.

T'other James came to us once a week on day release from school and in 1990 he joined the firm full-time, having to endure a year on YTS as I had done. I watched whilst in his early days with the firm James had to pass through the same rites of passage into the funeral profession that I had done and he too emerged from these trials unscathed.

In fact James went on to remain with the firm for 14 years, rising to become a fully-fledged funeral director and making the transition from the coffin workshop to a permanent berth front of house in the office. He also retained responsibility for the mortuary, an area in which he also had particular talent. All of this has a relevance to my story because much further down the line, long after I had left Thomas Broad & Son, t'other James and I would be reunited in competition with our old firm.

The other James' arrival at the firm in the late Summer of 1989 was the first significant event. The second event occurred just a couple of months later: my first murder – so to speak.

The row of cottages, The Boulevard, sat along an old raised pavement, at the top of Walkley Hill, in the Rodborough area of Stroud. The local pub sat at the top end of The Boulevard, on the corner by the junction with the other main road. Looking down the row of cottages it was not difficult to imagine how the scene must have looked years ago when they were originally built. But in modern times the view from the small, leaded windows of the cottages themselves was the parked cars opposite and the constant stream of traffic travelling up and down the hill.

Walls were built a lot thicker in the days when those cottages were put up and so, with only one property directly facing the row from the opposite side of the road, it was probably not difficult for a violent incident to occur inside Number One, down at the bottom end of The Boulevard and for someone to then leave the cottage without being noticed. Whoever it was that last visited Number One on that weekend in November 1989, holds the key to knowing how and why the occupant of the cottage, a 43 year old anorexic woman weighing just six stone, was bludgeoned and stabbed to death, before the cottage was set alight in an attempt to destroy the evidence.

The fire was still in its early stages when the emergency crews arrived, so the shocked firefighters who found the battered body in an upstairs room, as yet untouched by the flames, knew instantly that the woman who lay there had not been killed by smoke inhalation or any other effect of the fire.

It was Remembrance Sunday, 12th November 1989. Being a typical Sunday morning and being a typical nineteen year old, I was firmly ensconced in bed. My parents' house was just above a road junction and we had lived there long enough to no longer notice the constant stream of traffic that climbed the hill past our house on its way to the villages up on Rodborough and Minchinhampton commons. Likewise, we had got used to the sound of the sirens that the emergency vehicles always used as they approached the junction, negotiated the narrow one lane section of road outside my parents' house before racing on up the hill. On that Sunday morning I could hear the familiar sound of distant sirens climbing the first section of hill that lead up to the junction by the pub. But this time they went silent as they got closer. I didn't give it any further thought.

Later on (I won't say *how much* later on!) I got up and ambled downstairs. My mother was doing housework and I could hear my father returning home from walking the dog.

"The police have shut the road round the corner. I couldn't see what was going on, but there's a couple of fire engines and a load of police round by The Boulevard." My father said, as he came in, closely followed by Tibor, our Golden Labrador dog.

"I thought it was odd to hear the sirens stopping as they came up the hill. That must be where they went then." I replied, intrigued. "Hmm. I'm on call today so if it's anything serious I'll get to know."

We watched from the kitchen window, but the junction itself was as far as our line of sight would cover. The top end of The Boulevard was just around the corner, out of sight. Eventually curiosity got the better of my father and on the pretext of going to the post box just round the corner where the police cordon started, he wandered round. He was gone for twenty minutes or more and as he came back in my mother and I were eagerly awaiting whatever news he had gathered.

"There's been a fire in the cottage at the far end of The Boulevard. I was chatting to Simon from down the hill. He thinks someone's died in the fire and that the police are treating the death as suspicious."

A bolt of anticipation shot through me and my mind started racing. Our firm tended to do the majority of the coroner and police removals in the district, but here on the edge of the town itself it was a done deal that the call would come to us. I got washed and brushed up a little quicker than usual that morning, but I couldn't then settle to anything, instead silently willing the phone to ring. Father wandered round the corner again later in the morning and managed to speak to the police officer on the cordon, who confirmed that yes, there had been a fatality. He wouldn't say whether the death was in any way suspicious, but my father discerned from the level of activity that it was obvious the police were taking the whole incident very seriously.

Apart from one couple at the other end of The Boulevard, we didn't know any of the other residents of the row and certainly not the occupant of the end property where the fire was.

Calls to the business phones were diverted to Michael that weekend and the call finally came in at mid-day. Michael phoned me and said that the police wanted us to attend at a suspicious death at that address and we were to report to the scene at 1.00pm, to be ready to perform the removal the moment the Home Office pathologist had completed his initial examination. (The pathologist on duty, I would subsequently find out, was Professor Bernard Knight, who went on to gain fame just a few years later with his work on the Cromwell Street murders in Gloucester).

So that was it then, a potential murder! I thought as I put the phone down. Full of anticipation I made my way down to the funeral home to get the removal vehicle and meet Dewi, the bearer who was on call with me that day, before we made our way back up the hill. Dewi was himself a retired policeman from the South Wales valleys and this was all familiar stuff for him. We reported to the edge of the cordoned area at the appointed time, only to find that the police had been slightly over zealous in having us at the ready.

"We're just waiting for the pathologist to arrive, chaps." Said the officer on the cordon.

"Ok." I said. "Where's he coming from?"

"Cardiff."

Consequently it was an hour and a half hour later before we actually performed the removal itself. However, luck was on our side. The pub on the corner, next to which we were parked just outside the cordon, was a favourite Sunday lunchtime haunt of Paul, my boss. Having seen what was going on and having earlier chatted to my father, he waited long enough for Dewi and I to arrive and rather surprised us both by bringing out cans of coke and bags of crisps for us while we waited for the pathologist to finish examining the scene further down the road. We wouldn't normally have even considered consuming Coke and crisps whilst out on a removal of course, but as my boss was the one offering I was quite happy to accept – particularly as we had a long wait in store.

The police finally asked us to move down to the cottage at 2.30pm. Reversing the car down was also a signal for the local press reporters and photographers to gather themselves up and be ready to record the removal of the body. By this time word was rife that the death was suspicious and not simply an accidental fatality due to a house fire.

Paul had described to me, once or twice in the past, the murder scenes he had been to during his career – including one, just around the corner from this current incident, where the lodger in a large house had bludgeoned his partner to death with an ice axe. In a horrible twist of irony, the lodger's landlady, Pam Reeves, had since sold that property and moved into the bungalow right opposite the current murder scene. As we parked down in the middle of the cordoned area, between the cottage and Pam's bungalow, my mind was once again racing with anticipation. Pam was leaning on her garden gate and in a way that only she could, she called out,

"This is getting too much of habit for me James!"

We carried the stretcher up some steps that led from the road onto the raised pavement just in front of the cottage and positioned it by the front door. Then to my immediate dismay we were allowed no further than the front door step. Forensics officers in white overalls were crowded in the smoke blackened hallway and they handed the body – firmly enclosed in a body bag – out into our waiting hands. I don't mind admitting I was disappointed at the time, not being able to see anything, but with the benefit of further experience of suspicious death removals I wouldn't now expect anything different. Quite apart from the fact that cross-contamination of vital trace evidence can be prevented by reducing the number of people who have contact with the body, by also confining the number of people who even *see* the body, it immediately puts the spotlight on anyone later found to be talking about specific details of the body as someone who may be connected with the crime itself.

We delivered the body to the public mortuary at Gloucestershire Royal Hospital, where other police officers and forensics experts were awaiting its arrival. It would be two months before I would return there to remove the body back to our funeral home again for the funeral.

As neighbours in the vicinity my parents and I would later be routinely questioned by the police, asking if we had seen anything, or anyone, suspicious in the neighbourhood. Dewi and I were even called to the police station at a later stage, to have our fingerprints taken; there was one stray fingerprint at the scene that the police couldn't identify and they wanted to eliminate us from their enquiries. There would be photo-fits on the television and in the newspapers and eventually the case was featured in the BBC's Crimewatch programme too, but all to no avail. The case remains unsolved to this day.

As if bereavement by homicide is not unthinkable enough, the families of victims of a suspicious death can be denied the opportunity for an early funeral because of the need, in the interests of justice, to offer a defendant the opportunity to arrange an independent examination of the body. Although many homicides result in an early arrest, delays can be caused when there is no immediate suspect, so the coroner considers himself under obligation to retain the body - in some cases for a number of months - in the hope or expectation of charges being brought. This was the case with this particular murder, as there was not even a prime suspect. In the absence of a suspect, a second post mortem was performed by an independent pathologist for the benefit of any future defence case and after a couple of months the woman's body was finally released for burial.

For prolonged retention of bodies in such circumstances, public mortuaries have deep freeze facilities. At Gloucester mortuary there was one much larger fridge door at the end of the bank of fridges assigned for coroner's cases, marking the place of the three body capacity, deep freeze section. One of the difficulties with frozen bodies is that once they're removed from storage the onset of decomposition is much more rapid. You are in effect dealing with a biological time bomb, so once Paul had met with the murdered woman's family and the arrangements were made we agreed with the mortuary that we would remove the body just the day before the funeral.

When I brought the body back to our private mortuary and measured it for the coffin I was finally able to see this woman whose violent death was still keeping a police incident room open and whose assailant, or possibly *assailants*, posed such a mystery to identify. As part of the police's media appeals for information, a black and white photo of the dead woman had been released. It was a bizarre image – a photo more akin to something from a passport or even a police custody mug shot and her face stared out from the picture with a haunted, far away look. I remember having that unnerving, ghostly image in my mind when, with grim fascination, I had my first sight of that woman's face for real.

We none of us look our best when we die, but in her case the marks of violence had not faded of course and deep freezing had only made matters worse. The fact that she was a near neighbour, albeit it an unknown one, added a personal element to the moment, as well as to the rest of my minor involvement with the aftermath of her tragic death. Until such time as her assailant(s) may be identified, if ever, those of us who remember the case will always wonder why such a defenceless, frail woman was killed at all, let alone so brutally.

That was the first of eight murders with which I have so far come into contact with, either through carrying out the removal, handling the funeral arrangements or sometimes just preparing victims' bodies for other funeral directors. The next would be another female victim, this time found drowned in the boot of a car submerged in the River Severn - her husband was found guilty of the crime; then followed a man assaulted in a drugs related incident; a young woman battered to death by her husband; a young man stabbed after an altercation with another man during a night out on the town; a teenage girl stabbed to death with a Gurkha knife by her boyfriend in a sudden and frenzied attack whilst he was in a drunken stupor; a young man kicked to death by two men in an apparent homophobic attack and a caretaker attacked by an intruder in the grounds of a college.

One funeral I was involved with was notable not because of the victim, but rather the murderer himself. It all began eleven months before the Remembrance Day murder, in January 1989, when a local man due to give evidence in an armed robbery trial disappeared and his body was later found wrapped in tarpaulin and submerged in a lake in the Cotswold Water Park, near Cirencester.

Unlike the name suggests, the Cotswold Water Park isn't a conventional theme park but a rural area with over a hundred lakes created by allowing old gravel quarries to become filled with water. Spread over forty miles of countryside, the lakes within the park host a wide variety of recreational activities including sailing and fishing. Every year one or two bodies are recovered from the various lakes, usually following either accidental or suicidal drownings.

However, the discovery of the missing man's body in 1989 was a clear cut case of murder and the prime suspect for the murder was none other than the defendant in the armed robbery trial. His name became well known locally, but I shall refer to him simply as Mr. X. After a massive manhunt Mr. X. was found fifteen months later, having fled to Israel. He denied murder but in 1992, after what was referred to as the "supergrass trial" – a reference to the murdered trial witness, Mr. X. was found guilty and sentenced to life imprisonment.

Mr. X. did not admit to carrying out the murder until his own twenty year old son and another young man both died after the car they were travelling in crashed during a police pursuit in 2001. At the time I was carrying out hire work to the last remaining local builder/undertaker, who was conducting the convicted man's son's funeral and I was there on the funeral as the hearse driver. The circumstances of the police pursuit and subsequent crash itself were understandably mired in controversy and the fact that the young man's father had been convicted of a notorious local murder just added a further dimension to the whole tragic episode.

The funeral service took place in the parish church of a village nestling beneath the Cotswold escarpment, overlooking the Severn Vale.

When we arrived at the village church there was an enormous crowd gathered outside. The narrow lanes around the village were gridlocked and the scene was chaotic. Mr. X. arrived, handcuffed and flanked by prison officers. He begged to be able to touch his son's coffin one last time before it was carried into church.

However, despite the large police presence, together with all the press attention - all of which promised much newspaper coverage, the whole episode was, perhaps mercifully, overshadowed by international events of a kind no-one could ever have imagined. The day of the funeral was September 11th 2001.

Keith, my Father-in-Law, who was working as a bearer for me at the time, lived in the village where the funeral took place and he was on duty with me. Afterwards we had planned to use my hearse to travel on to our coffin supplier just over the border in Worcestershire to collect a specially ordered coffin. After the young man's funeral we called in at my parents'-in-law house for what was intended to be a quick toilet break and I phoned my wife Frances to check all was quiet back at the office before going to collect the coffin. Her first words were,

"Just switch on the television. Now!"

I was absolutely stunned by the pictures coming from New York showing the World Trade Centre ablaze following the terrorist attacks. I could imagine only too well the full horror of what I was seeing, because just over a year earlier, in May 2000, Frances and I had been in New York, visited the World Trade Centre and stood on top of the South Tower. Although most of the World Trade Center complex was off-limits to the public, the South Tower featured the Top of the World Trade Center Observatories on its 107th and 110th floors.

After passing through security checks added after the 1993 bomb attack, we boarded a turbo lift and were whisked to the 107th floor indoor observatory. Two short escalator rides up from the viewing area took us to an outdoor viewing platform on the 110th floor. It was a clear day when we went up there and the view, we were told, stretched for up to fifty miles. Then, months later on that fateful September day, like so many people around the world I couldn't believe what I was watching on the television and having been on top of one of those buildings, the attacks really did feel as if they had a personal element to them. As far as I am concerned September 11th is indeed a day that will live in infamy, for many reasons.

CHAPTER TEN
Over The Edge

"Has this fellow no feeling of his business, that he sings at grave-making? Custom hath made it in him a property of easiness."
Hamlet (William Shakespeare)

After I had been working at Thomas Broad & Son for about eighteen months they were offered the opportunity to buy-out another funeral director's business in the far south of the county, in the village of Charfield, on the border with what was then the county of Avon.

Clifford Boscombe had carried on the family business from his father and served an area from Charfield down to the outer edges of Bristol. This area took in a mix of both agricultural communities and the commuter belt. Clifford wanted to retire, he already had connections with my employers going back many years and with Charfield being only five miles or so from our existing branch office in Dursley, the acquisition seemed a natural one. Boscombe Funeral Services was a small business, conducting less than a hundred funerals a year, which in practice usually meant that nothing might come in for a week or two and then three or four funeral orders might all come in together.

The area Clifford Boscombe's little firm covered always felt to me like some kind of rural hinterland, sandwiched as it was between the Cotswold Edge and Bristol itself and all coloured by the peculiar reddish soil and stone of the area, so very different from the honey-coloured Cotswold stone I was more used to seeing. The presence of numerous stone quarries in the area also gave a hint as to what was to be found under foot. Indeed our contracted gravedigger encountered his very own rocky horror show in some of the churchyards he now inherited, one in particular where the graves could never be dug more than four or five feet deep because of the bed rock immediately below the soil.

Although the sleepy villages that Boscombe Funeral Services covered all seemed like the kind of places that you just drove through on the way to somewhere else, they would all generate a rich new crop of experiences for me.

In the early stages following the business acquisition, Michael went out on a few arrangement visits with Clifford to get a feel for how he did things. "Painfully slowly" was Michael's verdict. Clifford, an amiable little man with the look of a kindly grandfather, knew most of his client families personally and for him an arrangement visit was usually a relaxed and friendly affair. Michael, on the other hand, was used to an environment where we were doing ten times the number of funerals and the change of pace was very frustrating for him.

For me though, the foremost concern was rather more mundane, as I had to learn all the various formulas Clifford had for furnishing coffins: bar handles for men, ring handles for women, metal for burial, plastic for cremation, different types of bar handle depending on whether they were metal or plastic and so it went on. My colleague Rick meanwhile, had to learn a whole new round of hospitals and mortuaries in Bristol and surrounding areas. He left Thomas Broad & Son a little over six months later and his parting gift to me was all his notes regarding the routes, locations, opening times and individual quirks of each hospital and mortuary.

At first we were to carry on using Clifford Boscombe's office and chapel of rest, in a smart stone-fronted outbuilding in the grounds of his large detached house, situated directly opposite the long closed Charfield railway station. Just two hundred yards to the left of his house, spanning the railway line, was the road bridge that carried all road traffic through the village. I always found the sight of the abandoned station, which closed in the 1960's under Dr. Beeching's axe, strangely unnerving and I could easily imagine ghostly trains still stopping there in the dead of night. Ironically, there was an infinitely more dreadful reality connected with the place, but more than a year of visiting the newly acquired chapel of rest on the opposite side of the railway track would elapse before I would learn of the tragic events that had taken place by the station and the neighbouring road bridge.

Muriel, our secretary, was an absolute mine of information about our part of Gloucestershire, and Stroud in particular, along with virtually every family living there. She was born and bred in the area and had been with Thomas Broad & Son since 1970, employed by Thomas Broad II himself, at roughly the same time that Thomas II was priming his son-in-law Paul to take over the firm. Whenever a funeral came in Muriel would often start giving out a potted history of the bereaved family involved to anyone in the office with an ear to listen, but one day she brought a magazine article into work, handed it to me and said,

"Here love, this is something you might be interested in."

Muriel and I were alone in the office, so I took the opportunity to plonk myself down at Michael's desk and read about the Charfield Rail Disaster, together with the "lost" children and the woman in black. I was absolutely fascinated to learn that it had all happened right by the bridge just yards from Clifford Boscombe's house and that his father, as the village undertaker, had been involved in recovering the dead from the crash scene.

It happened on October 13th, 1928 and it was the arrival of a passenger train earlier than scheduled that led to sixteen deaths in a horrific blaze that gutted the passenger coaches. Passengers were travelling on the night mail train between Leeds and Bristol, steaming at sixty miles per hour towards Charfield, just as a goods train was shunting backwards into a siding. As the mail train hurtled towards Charfield, just another ten seconds needed to have elapsed and the goods train would have been clear, but it wasn't to be. For some reason never fully explained or understood, the mail train roared past a red signal and smashed into the goods train. The mail train landed on its side among the smashed wagons and three coaches behind the engine telescoped into each other, before falling against the road bridge immediately above the line. Gas used to illuminate the carriage interiors ignited and the massed wreckage became a blazing fireball.

Aroused by the noise of the crash, and the screaming of the panic-stricken passengers, villagers rushed down to the line. In a battle against the flames, villagers, railwaymen and passengers who had escaped unscathed made frantic attempts to free those trapped in the wreckage. The Railway Tavern, near the bridge (and still there today), was turned into a first-aid station, before its outbuilding was later commandeered as a temporary mortuary. More than thirty people were injured but it was much later before the grisly task of recovering the bodies could begin. Relatives later came to Charfield to identity them, but the charred remains could only be identified, for the most part, by rings, watches, cigarette cases, and, in one case by a piece of distinctive shirt the deceased had been wearing.

Allegedly among the dead were two children, so badly burnt that it was impossible to identify them. Yet no information emerged to link them with any of the other passengers; likewise nobody afterwards reported them missing or was able to provide any clue as to who they were with, or whether they were travelling on their own.

I was so absorbed in reading the article that I was only vaguely aware of Muriel tapping away at her typewriter and as I continued reading, a familiar name appeared on the page:

" "My Father had this old carpenter who actually put the dead into the coffins," recalls village undertaker Clifford Boscombe. "They were in full-sized coffins but that were only to ease the feelings of relatives, there were that little left of some of them. And he always swore he hadn't seen the bodies of the children." "

I could hear Clifford's voice in my mind as I read his words. The next part of the article had me laughing out loud. Muriel looked up from her typing.

"You reading the bit about Clifford appearing on local radio, love?" She asked with a knowing smile.

"Yeah," I giggled. "I can just imagine the journalist trying to understand Clifford's accent!"

" "I were on the radio once. Got up for seven o'clock, and all I got for it were a cup of coffee and a barge." This, it turned out, was a badge. Clifford says the questions are familiar

because, whoever asks them, they always lead in the same direction. He maintains that "My father had a very open mind on the mystery," "

It seemed from the article that Clifford's father was indeed right to keep an open mind, as the article went on to say that identification of the passengers' bodies proved so difficult that relatives of those listed as missing accepted the railway company's offer of a mass grave in the old village churchyard up on the hill. (I had been up to the old church before and wondered why the burials took place in the churchyard up there, when the actual funeral services always took place in the newer church down in the village itself. From reading the article I learnt all the local industry moved down into the valley in the 18th century, isolating the redundant church that overlooked the village, hence the presence of a second church in the village).

The article went on to describe the mass grave, saying that below the list of names on the memorial stone followed the words, 'Two Unknown'.

Apparently it was then said that, on every anniversary between 1929 to the 1950's, a mystery lady came to Charfield in a chauffeur-driven car to stand silently by the grave. A villager was quoted in 1978 as having seen the woman. "*...All I can tell you is that she was frail, always dressed in black, and came to the grave two or three times a year.The last time I saw her was sometime in the 1950s.*" he was quoted as saying. "*She always arrived in a chauffeur-driven limousine... She would put flowers on the grave and pray there. As far as I know, no-one ever spoke to her. She was elderly looking all those years ago, so I imagine she is dead now.*"

The identity of the woman is even now still a subject of fascination and debate and of course there are many who wonder if she was in some way connected to the "Two Unknown."

One Saturday morning not very long after reading the magazine article, I was up in the old churchyard at Charfield, to conduct a burial of ashes. Whilst I was waiting for the vicar and the family to arrive, a group of ramblers came wandering through the churchyard and asked where the railway victims' grave was. I took them over to it and gave them an impromptu history lesson on the disaster, leaving the hikers delighted that a local undertaker happened to be there in the churchyard and was able to relate the story to them. Revelling in my new role as impromptu village tour guide I didn't dare admit that the details were still fresh in my mind because I'd only learnt about the incident from the magazine article just a month or two earlier!

There is however an intriguing footnote to the story, courtesy of a second hand book I bought on a whim, nearly twenty one years after I first read about the disaster and the two mysterious children. Whilst browsing an internet auction site I chanced upon a copy of "Coroner – The biography of Sir Bentley Purchase" by Robert Jackson (Harrap, 1963). Bentley Purchase was HM Coroner for the St. Pancras district of London between 1932 - 1958 and was regarded as the foremost coroner of his era.

An arcane book like that would only have interest for someone like me anyway, but whilst reading through the well-thumbed pages I eventually reached Part Seven, luridly entitled "Murder, Fire And Children." One particular chapter in that section dealt with a fire at a Bloomsbury hotel on Good Friday, 1950, in which three people died. Two were identified, but the third body, although on examination found to be that of a woman, was charred beyond recognition and defied all attempts at identification. The coroner was forced to allow the body to be buried as "unknown" and it was the closing words of the chapter that caught me completely by surprise:

"Reluctantly Purchase closed the inquest with a verdict of accidental death and a promise to re-open the case if necessary. It was never re-opened, and Purchase asserted afterwards that the case was as strange as that, twenty years before, in which a boy of twelve and a girl aged five were killed in a train smash at Charfield, Gloucestershire, and never identified."

That is very specific information, considering there is no proof that the "Two Unknown" were even children. Did Purchase know something, or was it just another bit of idle rumour that somehow found its way from rural Charfield all the way to St. Pancras Coroner's Court in North London?

Unbeknown to Paul & Michael, the take-over of Clifford Boscombe's business had been watched with keen interest by Clifford's only competitor, Maurice Hynes, the funeral director in Wotton-Under-Edge

- the town three miles north of Charfield. After satisfying himself that Thomas Broad & Son had made the effort to take over Boscombe Funeral Services in a manner respectful of its heritage, Maurice too approached my employers with a view to selling his business.

Wotton-Under-Edge, the most southerly town in Gloucestershire, takes its name from its location immediately below the edge of the Cotswold escarpment. Situated on top of the Cotswold edge overlooking the town was the site of one of the early warning beacons used to warn England of the approach of the Spanish Armada in 1588.

Although geographically situated midway between Gloucester and Bristol, Wotton's whole orientation, like Charfield, is centred more towards Bristol than Gloucester. To illustrate the point, Maurice Hynes would later tell me that in his youth Wotton boys always married girls from Yate (just north of Bristol) because the hill was too steep to cycle up to Tetbury. Admittedly the hill, the "edge", is very steep, but nonetheless Tetbury is far more upmarket than Yate and I couldn't help but feel that the Wotton boys of old had been a little shortsighted in their romantic endeavours.

Whilst Clifford Boscombe had the appearance of a kindly grandfather, Maurice Hynes was the absolute epitome of an undertaker: tall but slightly stooped and with a long mournful face. The first time I set eyes on him was a year or two previously, at Gloucester Crematorium. Maurice was conducting a funeral whilst I was just calling in to deliver paperwork. He looked so much like an undertaker that I actually did a double-take as I drove past. Yet when I finally met him properly, that mournful countenance immediately lit up when he spoke and any last vestige of the Dickensian undertaker disappeared with the sound of his high-pitched, broad Gloucestershire voice. His favourite turn of phrase, as t'other James and I would never tire of secretly mimicking, was "Ooowah, now, best plan is...."

Maurice Hynes was, like Clifford Boscombe, another one man band, conducting around eighty five funerals a year. Despite their proximity, they each had their own distinct trading area and to their credit they suffered from none of the parochial jealousy that exists between so many funeral firms, be they large or small. They were, quite simply, business neighbours, as opposed to competitors.

Maurice's chapel of rest was also in the grounds of his house – this time built on to the rear of what had been his garage. There was a path down the side of the building, decorated with flower tubs, leading round to a rear door which took visiting relatives into a tiny hallway, through which the chapel of rest was then entered. Meanwhile Maurice would bring bodies in through the garage itself, still using an old fashioned "shell" (a coffin used solely for removals) over which a portable, electric coffin cooler he had bought years ago was then placed, to keep the body chilled.

Maurice's house was on a hill on one side of Wotton-Under-Edge and one window in his lounge looked directly across the valley basin in which the town sat, to the cemetery on the side of the opposite hill. Maurice's Wife Jean kept a pair of binoculars in the window sill and she told me that, knowing what time each funeral was scheduled to arrive at the cemetery, she could keep a close eye on the progress of a burial. If there was then an urgent call from a bereaved client she would be able to tell them how long it would be before Maurice could call them back. I suspect however, that it was more often just a method for Jean to judge when she would need to think about putting the lunch on.

Until the businesses of Messrs. Boscombe and Hynes were later amalgamated under one name and housed in proper premises in Wotton town, neither business had any mortuary storage facilities and so the usual procedure would involve me phoning ahead to the relevant hospital or mortuary in Bristol to ascertain a body size, before taking a ready-furnished, empty coffin down with me. That way I could then dress the coffined body in either of the Charfield or Wotton chapels of rest on the return journey.

I have never been keen on trusting mortuary technicians to measure bodies. Measuring a body properly can be a very subjective art and as I found to my cost at times, a mortuary technician's five feet nine inches by eighteen inches could be an undertaker's six feet by twenty inches. Often I would end up having to either wedge in, or pad out, mis-measured bodies into coffins.

On one memorable occasion, t'other James and I had travelled down to the former City Mortuary in Bristol to collect the body of a young woman who had been murdered by her husband, following a violent argument.

If ever there was a building suited to working with the dead, then the grade two listed building that originally housed the Bristol Coroner's Court was it. Even its location, in Stokes Croft, just outside the city centre, had a whacky reputation. Stokes Croft was a focus for art, music and independent shops in Bristol, together with three nightclubs such as The Croft, Lakota (next door to the old mortuary) and Blue Mountain. The area's character would later gave rise to a group of activists and artists calling themselves The People's Republic of Stokes Croft, seeking to revitalise the area through community action and public art. At least one local graffiti artist must have decided the gates to the mortuary yard needed revitalising too, because I always remember the gates sported a spray-painted cartoon face with the caption "Don't be a stiff, grow a quiff."

The Coroner's Court building dated from 1857 and was originally a Wesleyan Methodist school. Designed in Tudor Gothic Revival style it was converted for use as a coroner's court and mortuary in the mid 20th century. No doubt a very stylish example of architecture in its time, this gothic pile, complete with its arched windows and towers, sat very incongruously amongst the run down warehouses and modern city office blocks that comprised its immediate neighbourhood. The Coroner's Court and office occupied the upper floor whilst the City Mortuary was located on the ground floor.

My visits to the mortuary took place during the twilight years of its service, as it was closed down in 1991 after being classified as too outdated and expensive to run. Three technicians staffed the busy mortuary, which, as you entered, had an eight-table post mortem room stretching off to the right, a one-table isolation post mortem room on the left and then further in, the main body store housing fridges with capacity for approximately fifty bodies. It was all perfectly functional, but tatty and dated, with a very definite feeling that the mortuary probably hadn't changed much at all since it was first opened in the 1950's or 60's. However, the mortuary technician who normally dealt with the undertakers was such a character that I actually enjoyed going there to fetch bodies.

Eccentric or just plain batty, I could never quite make up my mind, but Ronnie was always friendly and like to chat… and chat…. and chat. He had wild grey hair, the thickest pair of jam jar glasses I had ever seen and with his white coat he looked like the archetypal mad scientist. When t'other James and I arrived on that particular day, the coroner's undertakers were also there, having just delivered one of the eleven hundred bodies the mortuary handled each year. Their presence, as two extra pairs of hands, would prove to be helpful.

Ronnie pulled the murdered woman's body out of the deep freeze section of the fridge and having positioned the lifter trolley next to the waiting coffin, he tilted the tray so that with t'other James and myself at one end and the coroner's undertakers at the other, we could slide the frozen body into the coffin. However, we quickly realised the coffin was nearly twelve inches too short for the body. I pointed out to Ronnie that I had taken careful note of the body size that one of his colleagues had previously given me on the phone and after some enquiries in the office it transpired I'd been given the woman's height when measured from heels to crown, as would be done during a post mortem. The problem was that whilst her actual vertical height would have been of help to a forensic pathologist, it bore no relation to the size I needed for the coffin, which, considering she was frozen hard, was whatever the length was from the tip of her pointed feet to the top of her head.

Fortunately, we always carried one stretcher in case a call-out came while we were out, so we placed the frozen body on that instead and on return to base we had to furnish another, larger coffin.

Unlike most places we went to, it was a requirement at the City Mortuary that funeral firms always sent two staff to remove bodies – to reduce manual handling risks for mortuary staff I presume. So it was then, that on another occasion t'other James and I had arrived for a removal, Ronnie asked us to firstly help him with a body which had not long been brought in: a young woman who had made a very botched job of hanging herself. There is a medical reality to suicidal hanging: asphyxiation is usually the cause of death, not a broken neck, as most people believe. Ronnie remarked that the woman's self-imposed departure from this life was a hard and painful one – a fact to which her contorted, congested features bore grim testimony.

With that bleak interlude over with and having then received and signed for the body we'd come for, we wheeled the coffin out to our vehicle. Ronnie followed us out into the yard, eager for a brief spell outside in the sunshine. On seeing that there were no other undertakers queuing behind us, Ronnie wanted us to stay and keep him company for a few minutes. James and I were left with little choice but to listen to him chattering away about this and that, before he abruptly announced,

"Course, as well as working here I've got a sideline." He flourished a business card he'd plucked from the pocket of his white coat.

I looked at the card and was momentarily lost for words. There was a black and white photo of Ronnie, still with his jam jar glasses on, dressed in a frilly shirt, wide-lapelled white jacket, flared trousers and white shoes. He was pictured in a really cheesy pose, with his left foot up on a block and his left elbow leaning on his knee. Across the top of the card was the caption: "Ronnie Bennett – International Cabaret Star." I genuinely didn't know whether it was a joke or if it was for real, but at that moment another private ambulance swung into the yard and to my huge disappointment I didn't get the chance to ask more.

Whether or not Ronnie was a real cabaret star or just a car ferry crooner, the City Mortuary itself had definitely enjoyed five minutes of fame as a filming location for the BBC drama series Casualty, which at that time was filmed in Bristol. There was a fake sign on the body store wall, referring to "Holby City Mortuary," a leftover prop reference to the drama's fictional city location. Even the coroner's undertakers had a bit part in the broadcast episode, stood waiting by their removal vehicle outside the mortuary doors.

Ironically, it was not the last time that the Bristol Coroner's Court and City Mortuary were to appear in the national media and television. In later years a legal dispute blew up with Bristol City Council in which the coroner was accused of acting improperly over the refurbishment of the replacement coroner's court, together with construction of a brand new, Home Office funded public mortuary at the new court complex, just south of the city. The legal storm resulted in the almost unprecedented suspension and later sacking of the coroner, after independent review by the Lord Chancellor and Lord Chief Justice. Meanwhile, at the time of writing the old court and mortuary building stands forlorn and empty, awaiting a buyer.

Long before the Bristol coroner became embroiled in the legal rumpus over updated facilities for his court/mortuary complex, I was having my own very minor battle with inappropriate working conditions. Once I had brought coffins back from the Bristol hospitals to Maurice Hynes' Wotton-Under-Edge premises, the only space I had to dress bodies in was in the garage, before wheeling them through into the adjacent chapel of rest. When I say garage, it was, like so many garages, used for everything other than keeping a car in. Many were the times I would be dressing coffined bodies whilst cramped in between the lawnmower, other garden implements and all the other tools and general garage junk cluttering up the place. Meanwhile, in the Boscombe premises at Charfield, there was only the chapel of rest for me to work in and if there happened to be more than one funeral on the go at any one time, then during viewings by relatives, the relevant coffin would be open for the family to see their loved one whilst the other, unrelated coffin would simply be covered with a pall.

I once went down to Wotton to deliver a coffin for a body that had previously been removed from a local nursing home. Maurice Hynes helped me to lift the body into the new coffin and then left me alone to set about dressing it. It was a baking hot day and when I had finished I knocked on the back door of the house to let Maurice & Jean know I was finished and that I was off on my way again. But Jean, bless her, said,

"Gosh, you look hot and bothered. Would you like a cold can of drink?"

I gratefully accepted and although I was wanted back at Stroud as soon as I was finished in Wotton, I decided head office could wait for me for another five minutes. I was absolutely parched and the can of fizzy limeade that Jean offered me looked so enticing, what with the condensation running down the side of it. I pulled the ring pull with great anticipation....only to find that the drink inside was frozen solid! I was too embarrassed to say anything to Jean so I politely fudged my way out of it and said "I'll be in trouble if I'm late back, so I'll take it with me, thank you."

My plan was to let it defrost on the way back to Stroud, which it did...eventually, but of course it was also warm by then. I was gutted. And still hot and bothered.

Most of the Bristol hospitals were easy enough to do removals from, but inevitably there was one exception to the rule. I won't say which hospital it was, but I absolutely dreaded going there. For a start, the access was bad. There was no space to turn in the yard area at the rear of the hospital where the mortuary was, so I had to reverse off a busy road into the yard entrance. Next I would have to do battle with the electric barrier. This involved walking across the yard to ring the mortuary doorbell, eventually getting a swipe card from the sloth-like technician when he deigned to appear, swiping the card on the barrier control, then rushing round to the driver's side of my vehicle and reversing in before the barrier started coming down again. Having done all of that I then had to face the dungeonesque mortuary and its troglodyte keepers.

There were two technicians – always dressed in casual civvies. The senior technician was unfriendly, off-handed and always mumbled his words, leaving me to try and decipher what he was saying, whilst his colleague just looked like he was out on licence from prison. Both of them had elevated surliness into an art form. The main hospital building was on a hill in the city centre and so the mortuary could either be described as on the ground floor, or in the basement, depending on your point of view. Basement was too polite a term in my view - I simply regarded that tatty, ageing mortuary as being a dungeon.

The Bristol hospital that I had to go to most often had three technicians working in its mortuary, two of whom were as good as gold, but once again the senior was an awkward, grumpy sod. However, our relationship improved somewhat after a near miss with mixed-up bodies.

I'd gone to the hospital on a routine removal and as I reversed my vehicle into the mortuary's ambulance bay I pulled up the handbrake, turned off the ignition and slumped in my seat for a few seconds, anticipating the prospect of having to deal with the grumpy technician. I reluctantly heaved myself out of the vehicle and wearily pressed the doorbell, hoping it would be one of the nice technicians who answered the door. No such luck, because, after a few minutes, misery guts himself appeared at the door. I wasted a polite "hello" on him and handed him the removal form. With a heavy heart I followed him in, past the collection of assorted redundant equipment that always made wheeling a stretcher trolley into the body store far more of a hassle than it needed to be. He opened up the mortuary register and ran a finger down the pages looking for the name on the release form, whilst I pulled the cover off the stretcher and began undoing the straps.

I waited patiently as he continued to study the mortuary register.

"Reynolds, William John." He pronounced at last, in the tone that you use when you have finally found something. Then, without him lifting his gaze from the register book, came a more deadpan pronouncement. "Oh shit."

He looked at the release form again, then back at the register, then back at the release form.

"Bollocks." He said.

I didn't say a word, I just stood there, waiting patiently, safe in the knowledge that if something was wrong then at least it couldn't be my fault.

He turned to me with a worried expression.

"There's two Reynolds in the book – Sidney Arthur and William John. Your guy's already been released," he announced, before telling me the name of the city funeral company who'd taken our body. I felt a sinking feeling in my stomach – but probably not anything as bad as the grumpy technican was feeling at this point. Both of us knew the ramifications. For him, the capital offence of releasing a wrong body. For me, the possibility that The Late Mr. Reynolds could already have been buried, or worse, cremated, by now. I admit I enjoyed seeing misery guts being left-footed by something, but that was before I knew the gravity of the problem.

"Shit." He exclaimed again, banging his fist on the register book. "We'd better phone them a bit bloody quick."

We? I thought. *It's your fault buster, not mine*. The technician beckoned me to follow him back round into the office where he quickly located the number from a list of their "regular" firms and the call was made. It was a very large company and it took a painfully long time for them to confirm, to our huge relief, that the funeral hadn't taken place yet. Then I thought, *hang on a minute, all we get from them is "the funeral hasn't taken place yet."*?! How about *"Oh God, we're sorry, we've got the wrong body!"* I had been in the profession for long enough by then to know that when body mix-ups occur, it isn't always just the fault of the mortuary technician, who normally has to take the blame, but often also because the funeral director's staff haven't checked the identity tags properly either.

I was even more unimpressed when the technician put the 'phone down and said,

"If you go down there now they'll hand the body over to you. They haven't got any staff or vehicles available at the moment but they've said they'll send someone up later to get the right body. It's our fault though, we've released the wrong Reynolds to them."

"And *they* didn't check they had the right Mr. Reynolds either." I said with some feeling, finishing the sentence for him.

As luck would have it, the firm in question had hosted my practical embalming examination just a couple of years previously, so I knew whereabouts in the city to find them. I drove straight round into their rear yard, wandered into the cavernous, but deserted garage and started to look for a member of staff. I headed towards the mortuary door, remembering the feelings with which I had last done that two years ago when I arrived for my embalming exam. At that moment I heard the slam of a car door outside and a few seconds later a chap in a business suit and carrying a briefcase walked into the garage. He had the air of a senior member of staff.

"You look lost. Can I help?"

"Hi. Yes. I'm in hot pursuit of a wrong body. Mr. Reynolds. There was a mix-up at the hospital. Two Reynolds's. You've taken ours by mistake."

"Oh right. Gosh, that's unfortunate. I'll go through and send someone out to help you."

With that he was gone. No expression of shock or even surprise. *If that was their attitude was it any wonder they had managed to take the wrong body?* I thought to myself.

I remained there in the garage for another five minutes before wondering if I should go round the front to the office myself, when a member of staff finally appeared. I recognised him from when I went there to do my embalming exam.

"Hiya. You come down for Mr. Reynolds then? He's in the holding room."

We stepped through the mortuary and into a large refrigerated room, identical to one I'd seen at the central Birmingham funeral home where I had studied on my embalming course originally. The bodies stored in this huge coldstore were all dressed and coffined, waiting for transfer either into one of the viewing chapels or to be loaded onto a hearse for the funeral itself. We threaded our way amongst the coffins, all sat on bier trolleys and the guy pointed to the nameplate on one of the coffins.

"Mr. Reynolds." He lifted off the lid and we both checked the wristband: William John Reynolds. The first thing that struck me was the irony of Mr. Reynolds being squeezed into a coffin that was very clearly too small for him. Maybe it was an omen that he shouldn't have been there in the first place. The other thing that struck me was that here he was, dressed in a standard blue taffeta gown, in a basic coffin and in amongst a roomful of other bodies all dressed and coffined in much the same way. He could so easily have been wheeled out of that room, loaded onto a waiting hearse and then driven away to be mourned over by somebody else's family, without anyone even being aware that his was the wrong body.

We extricated Mr. Reynolds from the coffin and I drove him back round to the hospital mortuary so that the – by now very sheepish – senior technician could check the wristbands for himself before he formally signed The Late Mr. Reynolds out of the mortuary register, which I also then signed, as the receiving undertaker. Considering all that had happened that afternoon, this final act of signing the mortuary register, a task which I have performed more times in the last twenty five years than I could possibly count, seemed really quite surreal. The technician realised the gravity of his mistake, but seeing that he was genuinely contrite I agreed that no purpose would be served by telling my employers and having the matter taken any further.

As I drove back up to base I wondered just how close Mr. Reynolds had come to being lost from his family. I have no doubt that old misery guts retreated into the mortuary office to ponder the same thought, because I found him far more amenable in future.

"Ooowah! 'Tis 'cus the one road runs into the other!" exclaimed Maurice Hynes, his doleful countenance lighting up with impish glee.

I was glad he saw the funny side. I personally took a dim view of one road running into another – no good ever came of it, as I was to discover one lunch time after being dispatched down to Wotton to do a house removal. I had one of the part timers, Dewi, a former policeman, with me.

We found the road we were looking for and Dewi swung our vehicle into the turning. We were in a small, T-shaped estate of 1950's semi-detached houses just on the edge of the town centre and as Dewi drove slowly along the road I began counting house numbers. We found number fifteen and as Dewi stopped I got out of the car, carrying a clipboard with a first call form and a company brochure folded inside. I knocked on the door and waited for a response as Dewi began reversing the removal vehicle onto the driveway, much to my irritation.

Even if there was space for us to park on the driveway, normal practice in the firm was to park outside the house while we had a preliminary chat with the family, before bringing the car onto the driveway to effect the removal. It was an unspoken rule but it always seemed presumptuous to park on a client's property until we were ready to effect the removal. I think that because death is an unwelcome intruder in any household, having to allow undertakers into your home as well just adds to the sense of invasion and as a company we were very sensitive to that.

However, as I was stood on the doorstep waiting for the deceased's wife to answer the door, there was little I could do to stop Dewi now and I gritted my teeth as he crunched the gears and generally made a hash of reversing on to the empty driveway. The woman opened the front door just in time to see Dewi reverse into her dustbin, which wasn't actually in the way anyway. *"The funeral director should always make his or her arrival discreet and unobtrusive when carrying out a removal from a private residence or nursing home"* was, I think, the phrase that was used in the Manual Of Funeral Directing, from which I had studied for my diploma.

Trying to smooth out the moment I smiled to the lady stood on the doorstep and said,

"Hello, Mrs. Browning. Please excuse our rather clumsy arrival." The lady looked at me with a slightly questioning look, but said nothing. There was a pregnant pause during which I expected her to at least say "Come in." After all, when she 'phoned the office she was assured that we'd be there as soon as we could.

"We're from Hynes and Boscombe, the funeral directors." I said, trying to move the conversation along a bit. I thought it was blindingly obvious who we were, but there wasn't even a flicker of recognition from her.

She broke her silence, but managed nothing more than a very rural sounding "Ah, is you?"

I ploughed on, reluctant to have to spell it out.

"Well...er, we're here to, er, remove your husband."

"Oohh, I see!" came the strangely worded reply, albeit at least in a tone which implied some recognition of our purpose at long last.

"Well, you can 'ave 'un if you want 'un, but 'im's in the front room watching telly at the moment!"

My stomach did an olympic grade back flip but my face could only manage a perplexed expression.

"Er, This is 15 West Bank? You're Mrs. Browning?" I enquired, in sudden panic.

"Ah! You want West Bank d'you?" she asked in a knowing tone. "T'is down that road there. This is 15 Bank Road! This is allus 'appenin' 'cus the way the road's laid out, see?" she pointed to the cul-de-sac forming one arm of the T-shaped estate.

"Oh no! I really am terribly sorry" I replied, "I'm so sorry to have disturbed you, it was a very stupid mistake for us to make."

"T'is alright, lovey. I think t'is quite funny. Wait till I tell 'im in there!" she guffawed, closing the door behind her.

Dewi thought it was hilarious and then took that as a cue to launch into another of his endless supply of stories about his time in the police force. I found myself wondering what the standard prison tariff was for strangling retired police officers.

It wasn't to be the only time I would have an eventful removal in Wotton with Dewi though. The next occasion was, once again, a house removal, this time from a rather less picturesque part of the town. After I had spoken to the family I went back outside, giving Dewi a nod that we were ready to begin the removal. As we carried the stretcher back into the house we passed the doorway to the cigarette smoke-fogged kitchen, where the family were gathered. The deceased had left a wife and two strapping sons.

"Oh christ. 'Ello Dewi! I didn't realise you were doin' this now!"

"Hallo Brian." Came Dewi's broad Welsh accent in response. "I saw the name on our ticket but I never made the connection. I'm sorry to have to see you like this. Hallo Daphne, ah, hallo Terry. 'Ave the old man been ill for long?"

"Hiya Dewi." Responded Terry, the other son. "Yeah, been ten months near enough. I'm glad in a way, it's been terrible for our Dad.... and for our Mum, watching 'im suffer." Their mother was lost in her own thoughts, tear-stained eyes staring at the floor, a cigarette poised between her quivering fingers.

The family insisted on watching us do the removal, which in itself was not unusual, but the deceased was quite a big chap and I would've preferred not to have an audience as I knew the removal was going to get a bit physical. The problem solved itself though when Dewi, without any hesitation, asked the two sons to each "grab a leg and help us lift." I would of couched it in slightly more tactful terms, but I was very grateful for their help and even more grateful for the fact that they would now be participating in the removal, rather than just watching us.

"That was a nice coincidence, you knowing the family." I remarked to Dewi as we drove away from the house. "Always nicer if the family think their loved one is amongst friends."

"Ah no, I ain't *friends* with 'em. No. I've known them two lads for years though, ever since I was stationed at Dursley, but only 'cus I've arrested the buggers a few times!"

CHAPTER ELEVEN

Declaration Of Independence

"If it be now, 'tis not to come; if it be not to come, it will be now; if it be not now, yet it will come: the readiness is all."
Hamlet (William Shakespeare)

Thomas Broad & Son had long become something of a monolith in the Stroud district. Their local brand was such that, if the name Cadburys meant chocolate, then by the same token if you lived in Stroud the name Thomas Broad meant funerals. Out of the collection of builder/undertakers locally, fate had decreed that Thomas Broad & Son would achieve pre-eminence and thus when the building side of the business was finally dropped, the firm had also been quick to recognize the future of funeral directing: smart, comfortable offices where arrangements could be made; a dignified private chapel of rest for families to see their loved ones; even proper refrigerated mortuary facilities like hospitals had.

Twenty four hour service was becoming the norm and undoubtedly the decision to forsake an answering machine in favour of ensuring the telephone was always answered personally made a huge difference to the company's success, despite huge cost to domestic and social lives in those pre-mobile phone years. The twenty four hour phone line also meant that the firm could be relied on by the police, who for many years would attend on behalf of the coroner at all sudden deaths. No matter which part of the district, if the police attended at a house where someone's Uncle Arthur had slumped into sudden oblivion at the meal table, or if any one of a hundred other possible scenarios of sudden death including accident, suicide or murder occurred, then nine times out of ten Thomas Broad & Son would be called.

By 1993 even I was getting very bored and frustrated in my current job. I had been at Thomas Broad & Son for six years as a full-time employee. I was twenty two years old, fully qualified and experienced but not feeling able to really explore my potential. Another funeral director had been brought in to the firm, a position I had assumed I was working up towards. The firm had been my universe for six years (seven, sort of) and up to that time I had been willingly indoctrinated with the idea, inevitable in a company of its size, that "everyone else does it different to us," as if Thomas Broad & Son's methods were the standard by which all else was to be measured.

But my attention was starting to turn outwards; I began to recognise that some things were actually done better in other firms. Most importantly of all, I began to realise that whilst Paul and Michael were wrapped up in their own myth, I no longer regarded their absolutes as my absolutes. I had entertained thoughts of working in another firm, but a sense of "out of the frying pan, into the fire" held me back. This hesitation allowed me time to feel the deep stirrings of a far greater and more radical ambition – to bypass alternative employment completely and instead branch out on my own. I'd come to see that the only way I could be fulfilled was to work for myself. This really surprised me – I'd always seen myself as a rise up the ranks type, like my father - a chartered accountant who rose from articled clerk to senior partner.

I had outlined my ambitions to my father on many occasions and we had actually looked at one or two opportunities further afield around the South Western region where funeral businesses were for sale. Then one evening my father handed me a set of property particulars from a local commercial estate agent. I looked through the paperwork and instantly recognised the property as being a redundant funeral home in Stroud. The building had housed the branch office of a well known national funeral chain, but the moribund branch had finally closed a year or two previously. My father and I had spoken before about the possibility of setting up a funeral company in the town, dominated as it was then, by my employers.

The empty property was located on a corner site, with the frontage on the main road, but with the side road climbing a slope up to the junction, so that the building was actually set over three storeys, with a three bedroom flat above, large shop front offices on street level and the ground floor containing the garage, mortuary and chapel of rest. Going to view the property one dark, drizzly Saturday afternoon was a fascinating experience. I had driven past the redundant funeral home so many times and wondered what it was like inside. The place was a time warp, having originally been converted from shop premises into a funeral home in 1974. By the looks of it there had not been any substantial investment in it since then.

I'd once had a random dream in which I'd moved to another part of the country and was working in a funeral home with a shop front office in a busy town. In the dream there was a castle up on a hill looming over the town, which could be seen from the window of the imaginary funeral home where I'd gone to work. Now, standing in the drab reception area of the real life closed down funeral home, I peered over the wooden partition behind the window display, with its dusty urns and plastic flowers. Just across the main road the houses opposite rose up steeply onto the hillside. Halfway up the hill, looming behind the houses, was the steeple of the local parish church, in just the same position as the castle had been in my dream. It was a vivid moment and I took it as a sign of fate.

The most interesting part of the tour was of course the mortuary area which, being windowless, was in pitch darkness and had to be viewed by torchlight. It was everything you would imagine the basement of an abandoned funeral parlour to be like: damp, cobwebby and still with the old mortuary fridge brooding in the corner. The estate agent told us he was the only member of staff who would conduct viewings on the property, his colleagues understandably put off by the thought of wandering round an old mortuary by torchlight.

To the estate agent's delight (and no doubt relief) we felt the building was perfect for our needs and within a relatively short period of time my father was able to use his contacts to locate investment funding for me to establish a new funeral business. After ascertaining that the current owners had no objection to another funeral director buying their former premises, the building was purchased. The building received a complete refurbishment and even the old fridge was found to be still in perfect working order.

The building had one interesting historic link: the main road on which it was located led out of Stroud into the Slad Valley, before passing through the village of the same name. Slad was the setting for the famous book "Cider With Rosie," Laurie Lee's autobiographical account of his childhood in the village. Mr. Lee later returned to live in Slad and he was a familiar face in and around Stroud. Paul, my now former boss, knew Mr. Lee and his wife well and Paul would often tell the story of how he was involved in the making of the 1971 film adaptation of "Cider With Rosie," filmed on location in the village. Paul was apparently hired by the film director to assist with the funeral scene. A little time after I had set up my new business I was lent a copy of the film on video and my fiancée Frances and I sat glued to our television screen waiting for the bit with the funeral.

I was disappointed to find that the scene simply showed the principal characters at the church, but there was no actual funeral directing action at all, let alone any sign of Paul. But what did appear later in the film, quite unexpectedly, was my new funeral home, albeit in its original guise as a Stroud Co-Operative Society grocery shop, to where Laurie Lee's mother, played by the actress Rosemary Leach, cycled on her weekly shopping expedition. I thought it was wonderfully ironic that Paul didn't appear in the film, but that my new funeral home did.

Paul would go on to conduct Laurie Lee's funeral in later years – the event being covered by a well-known national magazine, ironically this time with plenty of photo's in which Paul (and indeed my ex-colleagues) could clearly be seen.

I finally opened for business in August 1993, under a name taken from the neighbouring road: Lansdown Funeral Service Limited. I was doing all the funeral-related work, my mother worked unpaid as secretary/

receptionist/book-keeper, my father as unpaid accountant and also helping with call-outs, whilst my fiancée Frances also helped in many different ways. I can never thank Frances and my family enough for all that they did, and still do, to help me achieve my ambitions and support me in so many ways.

Paul and Michael were obviously far from happy, but nevertheless understood my ambitions. Of course, they didn't really see me as any great threat, such was the strength of their brand at the time. I always remember a good luck card that my parents gave me with a frog, wearing a crown and sitting on a lily pad on the front – "Now you're king of your own lily pad" they'd written inside.

The first adjustment I had to make was to the inactivity. I had detested the endless car washing and coffin fitting at Thomas Broad's, but now I missed the activity and the sense of being at the heart of what was going on. In effect I had to learn to slow down my professional metabolism. On the positive side however, I was now free to be the kind of funeral director that I wanted to be and the next seven years would represent an even steeper learning curve than that which I'd been through whilst training with Thomas Broad & Son. The main lesson I learnt, one which would radically change my mindset as a funeral director, was that there is a world of difference between working as a funeral director and actually running a funeral business. Over the next seven years I would have to confront the issue of bad debts, managing manpower and resources, as well as learning the hard way just how difficult it is to establish a reputation.

Those lessons were an important backdrop to my career at that time, but I was also enjoying the freedom to approach the process of arranging and conducting funerals in my own way; no longer with Paul and Michael looking over my shoulder and free of having to remain within the limits of the way they preferred things to be done. However, the old company line was more ingrained in me than I realised and it took a long while for me to shake it off and develop my own style.

One of my first aims was to resurrect the concept of wearing a top hat and paging (walking in front of) the cortege – a move which was unique in our area at the time and was very well received, despite Paul's dismissal of the concept when we had met on my first day of work experience seven years previously. It is strange nowadays that with so many people saying they want their funeral to be a colourful celebration of their life and not a grim, black-clad affair, that a surprising number, when they actually see the funeral director walking in front of the cortege, complete with the traditional top hat, really do appreciate the gesture. I think it's something so unusual to see that it gives the procession a sense of timelessness and gravity in an otherwise hectic, uncaring world. I think these traditions are far more ingrained in our culture than we might suppose.

Alternative-style coffins were the first sign of changing attitudes to funerals and my former employers didn't sell their first cardboard coffin until after I had left them in 1993. Certainly up until I left they would always offer a simple oak finished coffin unless the family expressed an interest in anything different, but in my new company I provided a photo brochure and encouraged families to make their own choices. The majority still opted for the simple oak finished style, which was fine, but I felt that at least I was giving them the choice from the outset. I also made it clear that my new company was open to the idea of arranging alternative styles of ceremony, even offering assistance to families who wished to carry out DIY funerals. Nowadays there are very few firms that don't offer similar help.

Being out on my own greatly improved my confidence, partly because of course the buck stopped with me, but also because I was maturing and developing in direct response to each new experience. By the time my second incarnation of self employment came a further seven years later, when I went on to purchase an already established company, I was comfortable with the unique approach I had created for myself.

At a practical level I was hiring the hearse and purchasing standard coffins from Fred Stevens Funeral Directors, run by William Stevens, in the nearby district town of Nailsworth. He used to hire limousines from my previous employers anyway and we got to know each other when he would bring bodies to me that needed embalming. William was sole proprietor of his well established family business and now I was self-employed we had a reciprocal arrangement for on-call and holiday cover and likewise we would provide each other with assistance whenever required. This reciprocal arrangement would bear much fruit years later.

Meanwhile, the funeral directors who had brought troublesome bodies to me at my previous employers now approached me direct for my assistance. I was also performing some freelance embalming work for various companies, either at their premises if they were suitably equipped, or by having the body brought to my new mortuary.

The die was cast then for the next three years, until a unique incident would herald a something of a diversification. The first year in my new business was of course a struggle and ended with a ten week hiatus during which we had no funeral orders in at all. Frances and I were by that stage married and she had taken over the small soft furnishings company where she used to work, moving the business into a separate part of our premises and thus the combination of funerals and soft furnishings work kept us afloat financially. In the second year we increased our number of funerals, gaining from not only our own name, but also with families who had used my funeral home's previous occupiers to arrange funerals with. But in the third year we discovered why the previous owners weren't too concerned about selling the premises to me, when they opened a replacement branch in the high street of Stonehouse, another small town within the district. By the end of year three we could see some impact on our funeral numbers. However, events in the autumn of that third year – 1996 – would alter my course somewhat anyway and as it turned out, prove extremely useful. That story is told in Chapter Thirteen, but there was more to happen before then.

It was during my fourth year in business, in the run-up to Christmas in fact, when one funeral in particular would prove to be something of a watershed moment for me; an intense experience that would give me a real insight into how I would need to be prepared to approach my work as a funeral director in future. A nineteen year old man lost his life in a tragic accident. I was contacted by his family, who lived along the road, on a Thursday lunchtime, just as Frances and I were preparing to go away for a long weekend. The young man's family were naturally anxious to see his body so, in my absence, William Stevens stepped into the breech and removed the body from the hospital – with some encouragement from the coroner's office and the mortuary staff, who I suspect knew what was coming.

Stroud is renowned for its thriving arts community and this young man and his family were very much at the centre of this community. Over the next ten days or so until his funeral, we had seemingly endless viewing visits from his family and many friends, often arriving in large groups. Rather than gather round the coffin and chat as most families do when they visit to see the body, all the people who came to see this young man would invariably sit in silence with the coffin for long periods, lost in their own thoughts and contemplations.

The period between taking the first call from the family, the nearly four hour long meeting I later had with them to discuss the funeral arrangements and the days then leading up to the funeral were all extremely intense and although it was a blessing that we had no other funerals to deal with during that time, I think the distraction might actually have been helpful. This was a situation where the normal conventions and expectations that had been ingrained into me in my time with my former employers just didn't apply.

This family didn't want tradition, it didn't mean anything to their son and consequently they felt it had no place at his funeral. I remember watching and listening as Paul & Michael at Thomas Broad's would be dealing with similar families and they would often find a way to talk them down into something far less imaginative and creative. At times I think they were worried about acceding to some requests because of "what people might think." My view was that if the request was that outlandish then people would realise that it must've been something the family asked for and consequently be impressed that we'd carried out those wishes. I vowed that when I was working on my own account I would do things differently and if a family wanted something more alternative, say a cardboard coffin or an alternative style ceremony, then I wouldn't hesitate to make it happen for them.

The funeral itself was a huge affair, with nearly five hundred people attending. It was held in the main hall of a local residential college and there were sound relays to two other large rooms where many other mourners were congregated. The young man had an open coffin during the non-religious ceremony. I had agreed with the family that it would be best if there was some music playing whilst I screwed down the coffin lid towards the end of the ceremony as it would take me a few minutes to complete the task and the music would provide something of a distraction, especially as I would have to screw down the lid in front of the whole congregation. The family came up with a suggestion which neatly incorporated something they had wanted to include in the ceremony anyway and so, as I closed the coffin, the entire congregation stood to sing the Beatles song "Let It Be" to the musical accompaniment of a member of the family. It was a truly moving moment. Burial followed in the town cemetery and after that nothing but a huge amount of relief for me!

As I said, it was a watershed moment. With the growing trend towards alternative and personalised funerals nowadays I will be forever grateful that I had that early and very comprehensive experience of this concept and it left me with much to consider about how I could accommodate such requests in future. There are practical issues, because such funerals can take on a life of their own if someone is not project-managing the whole event and yet that still has to be balanced with allowing ourselves enough leeway to enter into the spirit of the occasion, in order to work with the family and create something special.

No sooner was the young man's funeral completed than I was asked to help arrange a somewhat similar, though considerably less large, funeral for the mother-in-law of the sound system operator at the young man's funeral. The story of this second, very eventful and slippery burial is related in Chapter Twelve, but it demonstrates how funeral businesses grow; by the interconnectedness of people who will then come to us if they have been impressed by other funerals we have carried out.

"Oh, you did my aunt's funeral last year and now my husband has died. Can you do the funeral?" or "You did a friend of mine's funeral and it was really nice, so we've come to you." So your name and reputation grows. Obviously there are many other channels by which funerals come and it is the ongoing conglomeration of all these things that enable a business to become established and self-sustaining. One lady, after visiting different firms, chose us simply because, as she explained "I saw the picture of a fox that you had pinned on the board behind the reception desk. I took that as a sign because my husband loved wildlife." Her husband had taken his own life and maybe she had little else to rely on but things that she could take as signs or good omens.

One of the problems with starting out in business and building a reputation is the worry caused if something, even a very minor something, goes wrong. I'm not saying that now I run an established business that I don't care about things going wrong. In fact nothing could be further from the truth. But if a mistake or mishap does ever occur then obviously my immediate concern is to concentrate on taking appropriate action in response. No, what I mean is that back when running my first business, when I didn't have the distraction of another three or four funerals to get straight on with at any one time, I would end up stewing on even small hitches till they assumed a vastly disproportionate size in my mind. Take for instance the time the crematorium chapel attendant was making anxious faces at me as I led a procession in to the chapel, before he then started hissing to me

"The flowers! The flowers!"

He was a bit of a mickey-taker anyway and I thought he was just making fun of the rather tasteless flowers that had been chosen for the coffin on that particular funeral. That was until I heard a loud SPLAT! The attendant had been trying to warn me that the flowers were about to catch the door lintel of the chapel and when I heard the noise and looked round, there on the floor, behind the bearers and right at the feet of the family following, was the tasteless floral tribute – in a puddle of water and scattered petals. Miraculously, despite the scattered petals, the tribute as a whole was still relatively unscathed and with as much dignity and calm as I could muster, I picked up the fallen tribute and carried on leading the procession into the chapel.

Or the time I paged a cortege away from the house. As I reached the end of the road I heard a distant beep beep of a car horn and looked round behind the hearse to see that the limousine was still sitting outside the house. I marched briskly back down the road as the driver was gesticulating at me in some panic. The car was revving but not moving.

"There's something wrong with the car, it won't get into gear and move!" The driver complained heatedly.

Being more familiar with the limousine than he was I quietly asked the driver hop out and let me get behind the wheel so I could see for myself what the problem was, all the while painfully aware that the family were sitting patiently in the back whilst the hearse was waiting at the other end of the road. Being a vehicle with an automatic gearbox, I put my foot on the footbrake, engaged the gear lever into drive, took off the handbrake and.....the limousine moved slowly forward.

"How did you do that?" asked the puzzled driver.

"Get back in and try it yourself." I said.

The driver was rather rattled by now, as was I, but I calmly talked him through it.

"Put your foot on the footbrake" I said quietly, "engage drive, handbrake off"... and then the penny dropped.

"Oh god, I'm sorry James, "I didn't put it in gear. I forgot it was an automatic. I pushed the lever for first gear and of course that put it in park!"

I never did understand how he managed that, but it didn't matter at that moment – I was just relieved that the limousine was in perfect working order after all.

These examples were just silly, small mistakes; mistakes that undoubtedly detracted from the dignity of the funeral, but which nevertheless were innocent mishaps and nothing more than the price we pay for being human. That was certainly the view of the families involved with those two incidents I'm pleased to say. However, I have mentioned about keeping mishaps in perspective and how it is larger mistakes where we must look hard at what went wrong and deal with the consequences. I was soon to encounter an absolutely unprecedented situation where the gruesome consequences not only required me to take a stand against the perpetrator, but also prevent immeasurable distress being caused to the family involved.

I had been asked to arrange the funeral of an elderly man who had died in hospital in a distant county from Gloucestershire and who was to have a funeral service and burial at a church in the Stroud district. His death had been reported to the coroner and a post mortem was held. Once I received word from the coroner's office that the body had been released I made arrangements to collect it from the large district hospital in whose mortuary the body lay. I travelled up there myself and on arrival I found the mortuary technician to be uncommunicative and unhelpful. As I was a stranger in the camp anyway, I was only too glad to be in and out of that mortuary as soon as possible. In retrospect that was a mistake on my part, but as things turned out, a far greater mistake on the part of the technician. When I had transferred the body from the fridge tray onto my stretcher I had noticed something very strange about the post mortem incision on the head of the body, but I thought that whatever it was I would deal with it back in my funeral home. As I wheeled the body out and into my removal vehicle I could see another funeral director's private ambulance waiting, so I made a swift departure, thinking myself lucky that the wretched place wasn't one of my local mortuaries.

When I got back to base and unloaded the body I unwrapped the sheet to have a proper look. I was utterly gobsmacked by what I saw. After the initial shock of what I was looking at had faded, I started wondering if I had made a huge mistake in accepting the body from the mortuary in the first place. I was just so annoyed with the technician's attitude, coupled with being far out of my normal area, that I just wanted to get away again. But now I was kicking myself that I had come back all that way having already spotted something was amiss with the body before I took it. Thinking back to when I made the arrangements and how the deceased's daughter was adamant that she didn't want to see her father again, I wondered what would happen if she changed her mind and decided that she *did* want to see him after all?

The post mortem incisions had been sutured in the most careless way imaginable. The trunk incision, which runs from the lower neck down to the top of the pubis, was sewn in a manner that became progressively worse from the top of the trunk downwards, to the extent that by the bottom of the abdomen there was less than one suture per two inches of incision, making the incision disfiguring, leaky and loose. But the worst sight of all was the cranial incision, which normally runs from ear to ear across the crown of the head.

The fragile scalp of this elderly man's body was not so much incised as torn, with the jagged edges of the scalp stretched back across the still exposed skull cap with sutures that zigzagged across in exactly the same pattern as the strings on the side of a drum. I couldn't for the life of me understand how the scalp had been allowed to get into that state in the first place, let alone for a mortuary technician to think that he could get away with that kind of reconstruction. The skull cap was not re-seated properly after the post mortem either and so trying to stretch the fragile, torn scalp across it had just made matters worse.

There is something uniquely *dead* about a dead human body and as such, myself and my fellow morticians all the world over can handle bodies, put them in fridges, incise them, eviscerate them, suture them back up, embalm them and cosmetically restore them with no sense of revulsion or any lack of caring. We do this by being rational – by performing these procedures in the knowledge that they are the only methods we can use on a dead body; that these procedures are necessary to care for that body, but that they will in turn also provide a valuable service to the living.

But this time around, what I was looking at on the table in front of me seemed almost *abusive* in its carelessness. That dead human body, with no feeling or sensation, no consciousness, had still been *abused* through the utter carelessness of the mortician.

My first thought was to be thankful that the family (fortunately confined to just one daughter, a son-in-law and a grand-daughter) were adamant that they would not be wishing to visit the chapel of rest. I decided that nothing would be served by telling the family what had happened, but nevertheless I made my mind up to photographically record the damage to the body and lodge a formal complaint with the hospital.

This all happened at a time when internet access was still relatively novel, email was certainly not in wide use and digital cameras were only just starting to become available to the domestic market. Not owning a camera myself, I asked my long-suffering father to bring down his camera, one that had all the technical refinements that would allow us to take detailed shots of the body, and the two of us spent a less than congenial evening taking forensic-style photographs.

As I said, this was in the days when digital cameras were still the preserve of techno-types and gadget fans, so my next challenge was how to get the film developed. I could hardly take it down to a high street shop! Fortunately, I knew a chap locally who ran a photographic business and a discreet telephone conversation with him the following day revealed that one of his staff was an ex-police photographer who often now dealt with photographing injuries for insurance claims. Between them they agreed to develop the gruesome photographs in their own darkroom, thus ensuring total confidentiality.

Meanwhile I had re-sutured the body in a much more acceptable manner, but the damage was done. During that second day, when I had been arranging for the photos to be developed and working on restoring the body, I had also been mulling over my next move. I concluded that as a funeral director from one county, lodging a formal complaint against a large Hospital Trust in a distant county, about the condition of a body removed from their mortuary, there was the danger that my complaint would get fobbed off. I decided to take a different tack. I contacted my trade association and explained my predicament. They were appalled and immediately agreed to take up the matter on my behalf – particularly when I told them that I had photographic evidence to back up my complaint.

The photographs had been developed in duplicate and I kept one set lodged in my office safe. The other set were numbered and sent by Royal Mail special delivery to my trade association's head office. I also sent a comprehensive letter, under separate cover, setting out all the circumstances, identifying the deceased and referring to each of the photos by their reference numbers. In this way, if the photographs went astray, then no-one could identify the deceased from them, because the deceased's face was covered and the identity wrist band was also obscured with a piece of paper carrying a codename which was only translated in the separate letter.

Once everything had been safely received by my trade association I could do no more than wait. Despite days of anxiety on my part, the family stuck to their original intentions and did not request to visit the chapel of rest. The funeral itself, a small and very quiet affair, took place at the local church as planned. Then, less than a fortnight later, my trade association and I each received a copy of a letter from the Hospital Trust, responding to my formal complaint.

I really wasn't sure what to expect when I opened the envelope and saw the Hospital Trust letterhead. The letter was written not by the head of the Pathology Department, to whom my trade association had written, but the actual Chief Executive of the Hospital Trust. According to the letter, on receipt of the paperwork and photographs, the Head of Pathology had conferred with the mortuary manager and established that the mortuary technician who had carried out the botched post mortem was also the one who subsequently released the body to me. The rogue technician was a locum, covering another technician's scheduled leave. The letter went on to say that the mortuary manager and the Head of Pathology were so appalled by the incident and the photographic evidence that they felt the matter was too serious to be dealt with at departmental level and that it warranted action at Trust level.

The letter finished with a full and unreserved apology from the Chief Executive on behalf of the Trust and stated that the locum technician was of course banned from working in any of the Trust's mortuaries with a report of the incident also sent to the agency which supplied him. The photographs were safely returned to me and they were lodged in my office safe once again, whereupon I allowed a six month period to elapse, in case there were any further matters arising from the incident, before I destroyed every last photograph together with all the negatives.

My trade association were of course extremely pleased with the outcome and I remain sincerely grateful to them for ensuring that the matter was properly acknowledged and dealt with. The family concerned never knew, indeed still don't know, about that dreadful and unnecessary incident and all that happened as a result.

Self employment can be a very lonely place and I was deeply fortunate to have the support of my wife and parents. My father had spent nearly forty years helping his accountancy clients to run their businesses and I was hugely lucky to have all that knowledge on tap, whilst my wife Frances had that priceless ability to simply take everything in her stride and was never fazed by anything.

One very early lesson I learnt, or at least, learnt to start coming to terms with, was that I had to make a transition from simply doing the job to actually thinking of everything in business terms as well. I had to look beyond simply doing the best I could for every family and start discerning whether, just sometimes, they may have been taking advantage of my sympathy and willingness to please. I had to learn not only to identify what clients' needs were in terms of the funeral arrangements but also to look for any signs that they might not pay, either through a lack of funds or just out of a dishonest unwillingness to pay.

To begin with, this really shook my sense of altruism. It was a time of losing innocence and whilst now it's just another element of running a funeral business, back then it was a very hard lesson to learn. There were two bad debts during the time I ran my first business and looking back on both of those debts the warning signs were writ large from the very outset – if only I'd had the experience to spot them. But, as someone older and wiser once said to me, the man who hasn't made a mistake hasn't made anything and I learnt from those lessons; not just from the financial loss, but also learning to be hard when necessary and chase outstanding debts. That was not a nice thing to have to do and the resentment I felt about chasing the debts was equalled only by my resentment of the fact that these clients had put me in that position in the first place. I began to feel a great deal more sympathy towards my former employer Paul as I learnt for myself just what a difficult role being a business owner can be.

CHAPTER TWELVE
Down And... Out Again

"When you're alone you don't do much laughing."
P.G.Wodehouse

For most of the years I spent with my original employers at Thomas Broad & Son, there was a dusty, dilapidated, old church organ placed in a corner of the chapel of rest. It wasn't in working order, it certainly looked like it had seen better days and although the room was large enough to accommodate it, I always thought its presence added absolutely nothing to the tone of the chapel, in my view making the room feel like something from an old horror film – exactly what you *don't* want the place to feel like when bereaved families are visiting. However, the story of how the old organ actually came to be there is more interesting, as it occurred during my first experience of exhumation.

The very mention of exhumation tends to conjure up images of cordoned-off cemeteries at night-time, with police and forensic experts working under floodlight in white tents, in pursuit of evidence for a murder investigation. Indeed, watching scenes like this in television dramas or in the news is the only time that the public would ever see the process. Whilst dramatic scenes like that do occasionally happen, the reality is that exhumations are far more common than might be expected and the reasons are usually far more mundane. Most commonly exhumation takes place when the site of an old burial ground is to be redeveloped, or when relatives may, for various personal reasons, wish to have a deceased person reburied in another location.

My first exhumation came after Thomas Broad & Son were asked to arrange the disinterment of all the burials in front of a disused village chapel. Randwick Methodist Chapel had been closed some time previously and was due to undergo conversion into residential use. The main burial ground to the rear of the church was to be left undisturbed, but there were burials in front of the chapel which needed to be moved in readiness for the conversion work to begin. The village of Randwick itself sits high in the very westernmost folds of the Cotswold Hills, looking back across the Stroud Valleys, whilst behind the village a short ascent through the woods will take you to the crest of the hill and to panoramic views across the Severn flood-plain towards the Forest of Dean and the distant Welsh mountains behind.

Leaving aside the very rare exhumations carried out as part of police investigations, routine exhumations firstly require the appropriate form of authority. In all cases a Home Office Licence For Exhumation is required and if the body is buried in consecrated ground then application might also have to be made for an Ecclesiastical Faculty.

There will then be requirements from the local environmental health department, including for example having the site screened from public view and having a new coffin provided of sufficient size to accommodate the original coffin. (During exhumations, wherever possible the original coffin and its contents are lifted out of the grave intact and placed straight into a new 'shell' coffin in readiness for transportation to the place of reburial). In cases of exhumations from cemeteries still open to the public, there is often also a requirement that the work be carried out very early in the morning to ensure complete privacy.

In the case of the Methodist Chapel burial clearance however, the chapel was closed anyway and the graves were all long forgotten, so the early morning rule was dispensed with. Our gravedigger, who had been left in sole charge of the practical works, had instead erected wooden hoardings around the whole frontage of the chapel and simply carried on working through each day until all the buried remains were recovered. A final sweep was then made of the whole area in case there were any unrecorded graves that may have been missed. The exhumations were all from very old burials and as such consisted purely of skeletal remains. The remains were all placed in one communal coffin and one morning, my boss Paul, myself, the gravedigger, the environmental health officer and the local Methodist minister all gathered to conduct the reburial in a new grave at the rear of the chapel.

The main burial ground was on a steep slope behind the chapel and it overlooked a neighbouring house. No doubt the neighbours had long been used to having the redundant chapel and its overgrown graveyard looming over their garden; indeed they were probably more concerned with what any new buyer or developer would do with the building. But on that morning, as the minister led the way up the slope, it was rather funny to notice, out of the corner of my eye, the rather shocked look on the face of the neighbour, quietly pottering in his garden, as he looked up and suddenly saw four men, two in black suits, one in a tie and shirtsleeves and one grizzled and overall-clad, carrying a coffin up into the old burial ground!

Before the simple reburial, our gravedigger had kept the new coffin in the chapel and placed in each set of bones as he had recovered them. The chapel itself had largely been stripped out, although there was probably much still remaining that would have captured the interest of an architectural reclamation dealer. Unfortunately, there was one item that caught Paul's attention instead: the old chapel organ. He asked the gravedigger to recover the now useless instrument and within a day it had found its way into the chapel of rest at our funeral home, becoming a very large, very decrepit ornament. Whatever the logic for rescuing that old organ was, I never found out.

The process of exhuming bodies is complex enough, but burying the coffin in the first place can be equally complex, in the practical sense. As is typical of a hill and valley district like Stroud, all the soft soil is down in the valleys and all the hard rock is up on the hilltops. Unfortunately, many of our local churchyards and cemeteries are either on hilltops or hillsides, so not only is the digging hard and rocky, but the actual act of burial itself often requires a lot of balance, agility and strength on the part of the funeral director's staff.

We have one particular cemetery local to us which is so ridiculously steep that I have actually watched my bearers clutching the fence behind them with one hand whilst lowering the coffin with the other hand, simply letting the lowering webs slide through their fingers and hoping that they have enough strength in that one hand to allow a controlled lowering. In that same cemetery we have had to form human chains and use potholes in the turf to enable mourners to literally clamber up and down from the graveside.

One hillside burial, just a day before Christmas Eve, saw me very unceremoniously come a cropper. Directly across the valley from the village of Randwick, perched on another hillside, sits Rodborough Parish Churchyard. The few remaining plots for new graves at that time were located on a particularly steep slope at the end of the churchyard. This particular burial took place during a spell of very wet weather. To add to my concerns I had "family bearers", in other words members of family and close friends carrying the coffin themselves. As we neared the end of the path, ready to step onto the grassy slope leading down to the grave, I turned to the bearers and gave the word for them to lower the coffin from their shoulders and carry it "under arm," as we refer to it. I warned them to be especially mindful of conditions underfoot and told them I would stay in front of the coffin to act as a forward end brake.

Rodborough Churchyard is cursed not with rock but clay and having given instructions to the bearers I put one foot on the grass and stepped right on a patch of exposed clay. My legs went forward and out from under me and I was airborne for a split second before crashing down flat on my back. I tried to get up but of course my feet slipped on the same patch of clay again and back down I went. When I eventually managed to regain my balance my top hat had become shoved down over my eyes, I had grass hanging out of my sleeves, my previously immaculate frock coat was completely caked in mud and I was soaked through to the skin. We managed to complete the burial safely, once again forming a human chain to get mourners to and from the grave and the family's closing words to me were: "Mother was very theatrical and she would have loved the comedy of it all!"

Sometimes it isn't me, or the mourners, but altogether more elemental forces that can interfere with the smooth execution of a burial. I was self-employed by the time of the burial in question and the scene was All Saints Church, in the hillside village of Selsley, overlooking Stroud. Built in 1862, the church has a distinctive saddleback tower which the architect copied from a church in the Austrian Tyrol. All Saints is also a church famous for being at the very heart of the English Arts & Crafts Movement. The architect, George Bodley, was already acquainted with the Pre-Raphaelite Brotherhood of artists and it was his promise of commissions that contributed to the establishment of a fine arts design firm led by William Morris and other members of the Brotherhood: Edward Burne-Jones, Dante Gabriel Rossetti, Ford Madox Brown, and Philip Webb. This partnership of artists and intellectuals was called upon to design stained glass windows for the church, the most famous of which is the rose window above the west door, depicting scenes from the Creation that include a richly coloured Adam and Eve. It is considered to be one of the best small scale designs in stained glass by William Morris.

Unfortunately for me however, the afternoon sunshine that illuminates the William Morris window to best effect was conspicuous by its absence on the day when I buried a close neighbour of my parents in the churchyard. Of all the churchyards and cemeteries that we visit on business, Selsley is the most picturesque, with outstanding views over the basin of valleys within which Stroud sits. It's a wonderful location on a sunny day but when you have a full blown thunderstorm sitting over the five valleys, complete with torrential rain as well, the scene changes entirely. By the time the coffin was taken from the hearse and carried to the graveside the bearers and I were already soaked to the skin and the rain literally poured from the rim of my top hat every time I tilted my head.

As the vicar began reciting the words of committal I watched the sky turn ever blacker and literally the moment the coffin touched the bottom of the grave there was a bright flash, a crack of lightning overhead and the instant boom of thunder. If you were filming a horror film you could not have scripted the moment better.

On the other hand, had you been wanting to make a Carry On film then another burial in that same churchyard produced a scene any comedy film director would have been proud of: The weather was reasonable and everything was going to plan, except that behind the high wall that separated the churchyard from the neighbouring farmyard, something was clearly *not* going to plan. All the while the committal prayers were being spoken by the priest, his voice was being accompanied by the sound of a revving tractor, followed by loud bangs and shouts, then more tractor revving, more shouts, then louder revving, a very loud bang and then a lot of swearing....

The Christian gospels teach of Jesus' joyful resurrection three days after his death on the cross. However, in collaboration with a local authority I got this long-standing record down to three hours, albeit not under quite such joyful circumstances. The story starts with the death of Mrs. Cullen, an elderly woman whose funeral I subsequently conducted under Anglican (Church Of England) rites. She had converted to Catholicism in order to marry her husband, as in the far off days of her youth Catholics were not expected to marry outside of their faith. Mrs. Cullen reverted to her original denomination in death and thus was buried as an Anglican. The burial took place in a local cemetery, in a double-depth grave, to enable her husband to join her when his time came.

It was no more than a year later that Mr. Cullen's time came and he had a Catholic funeral, after which we buried him in the re-opened double grave. Everything had gone exactly to plan, which in this case was a particular relief as, somewhat unusually, there were two burials scheduled to take place on the same day at the cemetery in question. When arranging the funeral we had discussed with the family that we would need to time our arrangements around the other burial taking place that day, with the result that I agreed with the cemetery office that we would aim to be slightly early, whilst the other funeral firm agreed to arrive as late as they could, to ensure that our respective burials would not overlap each other.

After the burial was completed I hovered at a polite distance, available if anyone wanted me, but still giving the family some space and privacy to mingle together at the graveside before the limousine took them home again, where they had refreshments waiting. I glanced surreptitiously at my watch and decided I had another minute or two before I would need to gently encourage the mourners to depart and leave the cemetery clear for the funeral director with the second burial to arrive. Mr. Cullen's daughter wandered over to me and thanked me for making everything run so smoothly and, for the second time in twelve months, looking after the family so well. She turned towards the waiting limousine, hesitated and turned back to face me.

"There's just one problem, Mr. Baker. I think you've buried Father in the wrong grave."

There was no hint of emotion, no sign of anguish or upset in her facial expression, just a bland statement, as if she had just said something about how fortunate we were with the weather that day. I was used to the occasional outburst from bereaved clients when emotion momentarily overcame them. "I hate the flowers!" or "I hate crematorium services!" It takes the wind out of your sails for a second and then you realise that it's pent up emotion finding an outlet, popping like a cork in a bottle.

"I understand your concern, Mrs. Burford, but I can assure you everything's tightly administered and double-checked. It probably just seems different at the moment. Where other burials have taken place it's probably altered the general appearance of the area around the grave and now everything looks, well, different." It was the best I could think of to say at the time. I had 18 years less experience at that point. I would have countered it differently now.

"Oh." The daughter absently replied, half turning back towards the limousine. Then she turned back towards me again. "Well I hope it's the wrong grave, because since Mother died last year I've been tending that one over there." She pointed to another grave two rows back.

"Ah, right. Ok. Hmm." I replied, flummoxed, but not to the extent that I'd forgotten the other burial was due soon. "Clearly I'll need to look into this, then. Look, you've got everyone going back to your home, so may I suggest the car takes you back and meanwhile I'll speak to the cemetery staff and we'll get to the bottom of this. As soon as I find out what's happened, you'll be the very first to know."

The daughter accepted my promise without further question and all with the same unreadable facial expression that she had maintained throughout our conversation. Outwardly she seemed far less perturbed by the unfolding situation than I was. As the limousine pulled away I attracted the attention of one of the gravediggers, who was lurking discreetly behind a conifer tree. I repeated the conversation to him, to which he repeated all the assurances I had originally given the daughter, but in the end he agreed that we would need to make sure. The only way to do that would be to lift Mr. Cullen's newly buried coffin, clear aside the few inches of soil that covered the original coffin and look at the nameplate to see whether or not it was his wife's grave. We agreed that my team and I would vacate the burial ground and let the second funeral director arrive and carry on with his burial. Meanwhile the gravediggers would contact the cemetery office and explain the situation and we would then re-convene later to lift our coffin and make our investigation.

I had been in business on my own account for just over fourteen months and now here I was staring down the barrel of the kind of cock-up that every funeral director fears nearly as much as burying, or worse still, cremating a wrong body. (At least with burial they can be exhumed if, by some horrible chance, there is a mix up). Until I could get back to the office and pore over my paperwork I would have no idea as to how the mistake had occurred – if indeed there was a mistake – and more importantly ascertain whose fault it was.

About an hour or so later, I was back at the cemetery with the gravediggers and my father-in-law, who was one of my bearers back then. The gravediggers had already managed to lift Mr. Cullen's newly-buried coffin out of the grave themselves and it was hidden from sight under the folded Astroturf matting that the grave had been dressed with for the funeral. The foreman gravedigger announced that the original coffin lid had already rotted enough to cave in and judging by the smell coming up there was no way he was going to fish around inside there to locate the name plate. Before I could volunteer to look myself – I wasn't squeamish anyway and even less so if fishing around inside a mouldering coffin would tell us just whose grave it was – the gravedigger went on to say that it didn't matter anyway as he'd found a crucifix still attached to a bit of coffin lid that hadn't caved in. By his reckoning this meant it had to be the right grave because, pointing to the other coffin hidden under the Astroturf, he said "That one has a crucifix on the lid too."

My blood ran cold. "But when I buried Mrs. Cullen a year ago, I didn't put a crucifix on her coffin because, unlike her husband, she didn't have a Catholic funeral."

The gravedigger looked at me, a crestfallen look dawning on his face.

"Can I see the crucifix please?" I asked, in a tone far calmer than I was actually feeling.

"Yeah, I pulled it out for you to see" the gravedigger replied, picking up a mud-caked crucifix and handing the grubby artefact to me.

Another flush of adrenaline shot through me as I held the all-too-familiar item of coffin furniture in my hand. I felt as if I was handling an item of stolen property. Tearing my gaze away from this impromptu piece of contraband I tilted my head back up to face the expectant-looking gravedigger.

"I personally furnished over three thousand coffins for Thomas Broad & Son when I worked for them. I know their coffin furniture better than anyone. This is one of their eight inch, nickel plated, filigree crucifixes. I've put hundreds of these on their coffins for Catholic funerals. I probably put *this one* on originally."

I gently lobbed the crucifix back into the grave. The gravedigger, who I had known from my time with Thomas Broad & Son anyway, looked at me with an expression of resigned forboding. He picked up another small object that glinted in the sunlight and proffered that to me.

"Nickel plated, number three wreath-holder. Thomas Broad's only ever used Number Three's on burials. That's the wrong grave guys."

My statement was met with quietly muttered effing and blinding. We needed a plan – and quickly, so we agreed that I would discreetly remove the abortively buried coffin from the cemetery whilst the gravediggers would meanwhile report back to the cemetery office, before setting about opening the grave the daughter had pointed out earlier, to check *who* was buried in *that* one. Strictly speaking this was all illegal, as once the burial has taken place, removal of the coffin is classed as exhumation – a process with its own laws and regulations; but there are occasions when you have to apply common sense to a situation and in this case common sense demanded a quick and discreet resolution rather than observing proper regulations.

When we got back to my funeral home I left my father-in-law down in the garage hoovering the dirt off the unexpectedly returned coffin, whilst I waited for the telephone call from the cemetery office. They told me, to my great relief, that all my paperwork had been correct all along and that the mistake was theirs entirely. After the first burial a year ago, which had taken place on a Friday, another gravedigger – eager to finish work for the weekend – had decided to enter up the grave plot in the burial register on the Monday when he was back at work. Unsurprisingly his memory failed him on the Monday and he wrote in the wrong plot number. From that moment all our fates were sealed.

The correct grave had meanwhile been opened and this time the coffin lid had not caved in. Sure enough, staring back at us was the nameplate I had engraved a year earlier, with Mrs. Cullen's name on. We agreed that I would 'phone the family and explain the situation and that the burial authority would contact the family the following day with an official apology and an explanation of events. Still Mr. & Mrs. Cullen's daughter took the news of this development with a completely matter-of –fact attitude and whilst she was understandably not pleased, she betrayed no strong trace of anger or upset.

I have found in the years since, that many people either still think of the old-fashioned image of the often bumbling village undertaker, or they have no idea of what to expect at funerals anyway and so at times expectations are so low that people almost expect something to go wrong. The daughter did ask though, that as the priest had already conducted the burial, if I could just quietly rebury her father in the correct grave with no further fuss or ado, as soon as possible?

My father-in-law and I drove back to the cemetery again later that afternoon. After a quick look round to ensure that we were alone in the cemetery, the gravediggers, my father-in-law and I retrieved the coffin from my estate car removal vehicle (a hearse would have been too visible and indiscreet for our purpose this time around) and the four of us reverently lowered Mr. Cullen's coffin into his wife's grave. The deceased couple were re-united. Finally.

CHAPTER THIRTEEN
On The Beach

"Haphazardly scattered across the Indian Ocean, the mysterious, outrageous and enchanting Comoros islands are the kind of place you go to just drop off the planet for a while."

Lonely Planet Travel Guide

I had an ambition to enter the funeral profession, which I was able to achieve. As time went on I then felt an ambition to have my own business. I was able to achieve that too. But still I wanted to experience all that the profession had to offer and to be as well qualified as I could be – in all senses of the term. Becoming self employed in 1993 would give me the freedom, three years later, to fulfill one very particular and unique ambition.

I've already mentioned the Charfield Rail Disaster and an often overlooked aspect of the funeral profession in Britain is its integral involvement with the response to mass fatality incidents around the world. This is nowadays referred to as "Disaster Management" and there are international companies specialising in this area of work.

Professional disaster management in the UK and indeed probably the world, traces its origins right back to 1906, when J.H. Kenyon Ltd., a well respected London funeral company, already renowned for their work as pioneers of both modern embalming and international repatriation work, were asked to embalm and repatriate twenty eight fatalities, nearly all Americans, from a rail disaster in Salisbury in Wiltshire. Kenyons were then involved with their first air disaster in November

1929, when an Imperial Airways Junkers monoplane crashed in England. From that point on, their company was called upon to respond to many more disasters as the civil aviation business grew. To meet this growing demand, a separate disaster management division of the company – Kenyon Emergency Services - was formed in 1975. I was fascinated by the work of this team and disaster management as a whole.

A chance conversation during a South West regional meeting of the British Institute Of Embalmers early in 1996 led to me applying to join a newly-formed disaster response company, similar to Kenyons, who were recruiting various mortuary and funeral professionals, embalmers included, for their field team. A few months later I attended an induction day for the recently formed team at a funeral home in the Midlands. I was surrounded by a mixture of people who had seen active service on other incidents during what one embalmer referred to rather pointedly as the halcyon days of disaster management in the late 1980's, when one high profile disaster seemed to follow another; the Zeebrugge ferry sinking in 1987, the Piper Alpha Oil Rig explosion and the Lockerbie bombing in 1988 and the Hillsborough football stadium disaster in 1989. One team member had even been responsible for co-ordinating the mortuary response following the desperately tragic Aberfan mining disaster in 1966. At a later stage he was kind enough to share his experiences with me in great detail and even thirty years later the emotion in his voice was palpable. The conversation left a deep impression on me.

Much of the induction day was filled with the full-time team leaders explaining the structure of the team, the procedures to be enacted following a request for assistance at an incident and the basic logistical plans for mobilising the team and maintaining it during a deployment. It was a grimly absorbing day for me and at the end of the induction all the team members dispersed back out onto various motorways home, knowing that the next time we would see each other would be for the real thing.

The months passed and summer came and went, all without incident. Then on 12th November 1996 a mid-air collision between a Saudi Arabian Airlines flight and an aircraft from Kazakhstan Airlines occurred. All 349 people on board both flights were killed, making it the deadliest mid-air collision in history. I received a telephone call from one of the team leaders warning me that the team was being put on forty eight hour standby to respond, but less than twenty four hours later we were all stood down again as it appeared the authorities had their own plans under way.

I settled back into my normal routine and although I spent some days wondering what the incident would have been like to work on I soon put it out of my mind, knowing that sooner or later my baptism would come. Sooner, as it happened. It was a Sunday teatime, just twelve days later and I was sat in our flat above my funeral home waiting for the news to start on television. The newsreader read out the main headline and in the corner of the screen was a photo of what looked like an aircraft tail fin sticking out of the sea, like a huge red, green and yellow shark fin. Below it on the screen was the word "Hijack". I stared at the TV screen, fixated.

The newsreader explained that a hijacked Ethiopian Airlines Boeing 767 aircraft, en route to Nairobi, had crashed into the Indian ocean. The crash had happened just a few hundred metres off the coast of a holiday resort on the island of Grand Comoro and the tail fin photograph had been taken from the beach. The television images showed casualties and bodies being carried up the beach, before cutting to dramatic footage of the aircraft's final, fatal descent into the water, captured on video by a South African couple on honeymoon, who witnessed the crash as they sat on the beach. The newlyweds began filming the low-flying plane because they thought it was part of an air show for tourists.

Ethiopian Airlines Flight ET961, carrying 174 passengers and crew from over twenty five nations, was en route from Addis Ababa, Ethiopia, to Nairobi, Kenya, when it was hijacked by three Ethiopian men. The accepted course of events seemed to be that the three hijackers forced their way into the cockpit, armed with a fire extinguisher and small fire axe, and they threatened to blow up the plane with a bomb. Apparently it was later discovered that their "bomb" was actually just a covered, unopened bottle of liquor.

The hijackers announced on the intercom that they were opponents of the Ethiopian government seeking political asylum and had recently been released from prison. They said they were changing the direction of the aircraft and threatened to blow it up if interfered with. The hijackers instructed the flight crew to fly to Australia, saying they knew that the aircraft could reach that destination since a Boeing 767 could fly for eleven hours non-stop. During the course of the hijacking, however, arguments erupted as the captain tried to convince the three men that the plane was running out of fuel. According to survivors' accounts, the passengers had the impression that the hijackers were unprepared and not well rehearsed. The consensus among the passengers was that an assault against the hijackers would be safer when the aircraft landed for refuelling rather than in flight because of the risk of provoking the three men to detonate the bomb they claimed to possess. Overall, the passengers remained calm; however, they were unaware of arguments between the captain and the hijackers regarding the aircraft's destination and dwindling fuel reserves.

The captain was flying south along the coast of Africa instead of east over the Indian Ocean toward Australia, as instructed by the hijackers. Three and a half hours into the flight, one engine ran out of fuel and stopped, causing the aircraft to drop from 39,000 to 25,000 feet. Upon realizing their instructions had not been followed, the hijackers reacted furiously and threatened the flight crew, who in turn thought the hijackers would detonate their "bomb." According to survivors' accounts, the captain then made his only communication to the passengers during the ordeal, informing them of the loss of one engine and the imminent loss of the other engine before giving instructions to prepare for an emergency landing. The survivors said the captain ended his final announcement with the words "Do what you want to the hijackers."

The plane continued to lose altitude and began to sway. Much of this time was consumed by the cabin crew giving instructions to the terrified passengers about the proper use of life jackets. According to a survivor's testimony, despite the crew's instructions, sounds of life jackets being inflated could be heard throughout the cabin. One survivor went on to tell of a young Kenyan man, travelling with his three children, holding them tightly in his lap, their lifejackets strapped on tightly and already inflated; the hysterical children's faces contorted with fear whilst their father's face wore only an expression of grim resignation, two single tracks of tears running down either side of his face.

The aircraft was approaching the Comoros Islands. The pilot had been given clearance to land at Moroni Airport, Grand Comoro, but he knew the aircraft would not reach it; instead he tried to land the plane in the water near the Galawa Beach resort. The hijackers, however, realized their plan had failed and attempted to take the controls. As the aircraft approached the water a wing tip skimmed the water, causing the plane to overturn at least once before breaking into three segments. The plane crashed five hundred yards from the resort's beach and sixteen miles from Moroni Airport. Out of the 174 passengers and crew, 127 of them lost their lives.

The majority of the survivors were hanging on to the fuselage section, which was floating, but the rear section of the plane was submerged, hence the tail fin later being seen pointing out of the water. The fatalities ocurred either as the result of the initial impact, or from subsequent drowning because the passengers' inflated life jackets prevented them from swimming out of the water-filled fuselage. The pilot and copilot survived but the hijackers did not.

What followed was what could only be described as one of the most unlikely rescue attempts in civil aviation history, as staff and guests from the beach resort hotel used small pleasure boats and even pedaloes to reach the crash site. Utterly bizarre scenes ensued as passengers who survived the initial impact found themselves floating in debris-strewn sea water, half blinded by the spilt aviation fuel in the water, before seeing a tourist-led rescue force bearing down on them in pedaloes...

The Galawa Beach Hotel's open air restaurant was quickly turned into a makeshift triage station, staffed by some French and South African doctors on holiday at the resort, before the survivors were later sent to the island's woefully ill-equipped Moroni Hospital. Although in the previous five years ten hijackings had taken place in Ethiopia, all by Ethiopians seeking political asylum and escape from conditions of poverty in the country, Flight ET961 was (until September 11th 2001) one of the deadliest hijackings in civil aviation history.

My team leader phoned later that night and told me the team had been engaged by the airline's insurers and that this time around there was no doubt that we would be going. Further instructions were to follow regarding the travel arrangements. After a surprisingly long delay of a few days, the travel instructions finally came through and I duly reported to a hotel at Heathrow Airport, where the team had been asked to assemble. After gathering for a briefing in the hotel lounge we took it in turns to troop upstairs to the team doctor's room to have the necessary inoculation jabs and anti-malarial tablets. I personally then spent a long and miserable night in the bathroom of my hotel room suffering with side-effects from the tablets.

Before we boarded our flight the following morning we were told the principal team leaders had travelled on ahead to assess the situation and set about arranging accommodation and establishing temporary mortuary facilities for us to work in. We were also told a little more about where we were going:

The four volcanic Comoros Islands were located in the Indian ocean, sandwiched between Mozambique and the island of Madagascar. The Comoros islands, originally French colonies, were politically unstable, experiencing almost twenty coups since independence in 1975, earning them the clever but worrying nickname "Cloud Coup-Coup land." Apparently it was only a half joke to say that a Comorian president was lucky if there was time for his official portrait photograph to be taken before armed rebels were once again stood at his office door. Grand Comoro, where we were headed, was the largest island and on which Moroni, the capital of the island group, was situated.

We had a fourteen hour flight down there, the drudgery of which was alleviated somewhat by being in business class, on an Air France Boeing 747 aircraft. That was to be the last time of comfort and sanity for ten days and I spent all of the flight trying to imagine what might be awaiting us. As we approached the islands I stared in fascination at the black, jagged volcanic landscape below. The runway of the tiny island airport was barely large enough to take a Boeing 747 and our landing was correspondingly brief and to the point.

When the hatch opened a blast of hot tropical air rushed into the cabin and the moment I stepped out into the glare of tropical sun I suddenly found it very hard to breathe in the cloying equatorial heat. As if adjusting to the unbearable heat wasn't enough, I was also instantly aware of a crowd of native women, stood on the roof of the dilapidated terminal building, all hollering with a peculiar high pitched call, as if engaged in some traditional tribal greeting – or, as I was more inclined to imagine, some cannibal ritual to celebrate the arrival of the unsuspecting white man. I apologise if that sounds a little racist, but there I was, thousands of miles from home, in a politically unstable little banana republic island, barely able to catch my breath in the blast furnace heat and now being bombasted by a tribe of screeching, wailing native women. I decided there would be safety in numbers so I stuck with the rest of the team like a limpet.

We were met by two of the advance team members in the chaotic arrivals room – "terminal" was rapidly becoming far too grand a term for the building. A few tense moments passed as our passports were examined by threatening looking men in military uniforms and then, having fended off natives eager to earn tourist currency by carrying our luggage – although probably not to where we wanted it to go, we emerged out into the dusty surroundings of the island airport.

Being a born fatalist I was already beginning to wonder whether I might ever see home again. I watched with growing unease as up to nine or ten people would pile onto dilapidated pick-up trucks and trundle out of the airport, but to my relief we had an elderly but serviceable bus waiting for us. It was decided that as time was of the essence, rather than waste half a day going to the hotel first, we would instead travel directly to the temporary mortuary in the north of the island, in a small town called Mitsamiouli. Meanwhile, the crash site itself, off Le Galawa beach, was ordinarily an hour's drive away from the temporary mortuary.

The Comoros Islands are very poor and underdeveloped and the sight of real poverty, with people living in shanty villages of tin shacks at the roadside was a real shock to my cosseted, first world sensibilities. The journey in the bus was bone-jarringly bumpy and at times downright petrifying, as what few other vehicles there were on the island would either overtake in dangerous places or else the bus would have to swerve to avoid other drivers approaching us on whatever side of the road took their fancy.

As if our journey wasn't proving to be dangerous enough already, I was acutely aware that the last military coup had only been staged a year or two previously and my nervousness about the entire expedition was growing all the time, compounded by the knowledge that it was too late now anyway, there was no going back. As I sat there clinging to my seat, pondering my predicament, the sheer lunacy of it all was offset by the strange juxtaposition of views from the windows of the bus: to one side the Indian Ocean, azure blue and glinting in the blazing sun just like it would in the glossiest of holiday brochures, whilst to the other side rose the enormous Karthala volcano, brooding over the island in all its lava-black, glowering majesty.

We arrived in the town of Mitsamiouli, passing heaps of rubbish at the roadside where animals and even a few humans were routing through the stinking piles of waste, past run down but obviously ex-colonial buildings between which locals loitered in whatever shade they could find. It wasn't difficult to see that some of the pockmarks in the shabby, faded walls were actually bullet holes.

The bus turned into a compound and as soon as we drove in the gate was closed and locked behind us. I felt sure I saw an armed guard hovering by the gate. As we got out of the bus the third and final member of the advance team was stood there waiting for us. He looked sunburnt, sweaty and utterly exhausted.

"Welcome aboard everyone. It's hell on wheels here."

Wherever my comfort zone might normally have been at that time, I was in the furthest place from it and this greeting reinforced the feeling further. We were in the yard of a small, industrial-type compound. Considering the all too obvious lack of infrastructure on the island, the advance team had miraculously discovered that the crash victim's bodies had been moved to a refrigerated storage facility within a warehouse. The facility was designed for food storage and certainly not designed or intended for mortuary use. But nonetheless it was a fridge and that was a rare blessing on this backward outpost of humanity far out in the Indian Ocean.

We were told to retrieve our mortuary scrub suits and wellie boots from our luggage, before leaving the rest of our baggage on the bus. For the rest of the day we began the process of getting the temporary mortuary up and running. Each body was intended to undergo a four-stage process: 1) Retrieval from the fridge, number coding of the body and listing of clothes, any jewellery and identifying marks such as tattoos, etc., 2) Examination of the teeth by the team's two forensic odontologists, ready for matching with dental records that the home team would be gathering from relatives, 3) embalming and 4) placing in coffins ready for repatriation from the island.

It would take another day or two before everything was finally set up in a way that enabled the bodies to be processed in an efficient flow, but even then such facilities as we could muster would be hopelessly inadequate for the task. Because of legal & insurance formalities it was nearly a week after the crash before we even arrived on the island and we would be faced with over a hundred bodies, all bearing the appalling effects of immersion in sea water, coupled with exposure to humid tropical heat and of course the effects of the crash impact itself. "Hell on wheels" was just about the greatest understatement anyone could have made.

Having been flying for fourteen hours, landing in some strange tropical country, then taking a bumpy, hair raising road journey to a barely serviceable temporary mortuary was enough to take in. But now, stood in a dingy industrial warehouse under the sickly, weak illumination of an inadequate number of flourescent lights, I found myself staring at a very badly deteriorated, swollen body, barely able to draw breath because of both the stench and the air humidity, and it suddenly hit me just what I had dropped myself into.

I was thousands of miles from home, on an isolated little island the rest of the world barely knew existed, where social unrest could kick off at any time and now faced with the thought that every dead body I would handle would be as bad as the one in front of me. My experience of embalming difficult and decomposed cases was limited and I had made the mistake of assuming that because team procedure required that embalming would be carried out by pairs of embalmers on each body, I would have the opportunity to learn and expand my skills "on the job," so to speak. But in those first hours I realised this wasn't going to be the time or place for trainees. The pressure was on and there would be no respite, indeed no escape. It was no exaggeration to say that I had never felt so alone and helpless in all my life.

Late that evening, to my eternal relief, we finally got away from that hell hole and back to our hotel – ironically the Galawa Beach Hotel, off whose private beach the crash had originally occurred. We were all so tired on that first night that after catching the tail end of the evening meal in the open air restaurant, we all quickly retired to our rooms, where in my case I soon found out I would be sharing. A tiny movement down on the floor had caught my eye and I looked down to find a small lizard fixing me with an unblinking stare, of the kind that implied he knew as well as I did that one of us wasn't really welcome in the hotel room at that moment. I chased the creature round the room with a shoe in my hand, but he was too quick for me and he eventually disappeared unharmed.

The change in surroundings from temporary mortuary to luxury beach resort hotel full of tourists made the whole experience even more bizarre. I had a restless night, disturbed by all the thoughts spinning in my mind, all set to a soundtrack of constant humming from the ineffectual air conditioning unit in my room. Finally I gave up trying to sleep and just sat there in the dark, with the balcony doors open in a vain attempt to let some fresh air in. I could hear the sound of waves very close by and wondered just how near the sea actually was. I tried to see out into the darkness but all I could make out was a vague, but very large, white shape somewhere out there beyond my window. With sleep barely an option I was instead eager for daylight, just to be able to get my bearings and see what was out there.

The next morning dawned with bright sunshine and pure blue skies. I got out of bed and in doing so realised that somehow I must have got off to sleep after all. I instinctively made straight for the balcony doors to see what was outside. My first floor room was on the edge of the beach – literally. The beach itself was everything you would expect a picture postcard tropical beach to be and there was that pure blue sea again, shimmering in the sunlight. But all these things were reduced to incidental details at that moment, because my eyes were glued to what had been the ghostly white object I had noticed the night before.

There on the shoreline, just forty yards from my window, was a section of fuselage from the aircraft. It was upside down, with the remains of one wing stretched out and wedged into the sand. Through a partial curtain of severed wires, cables and steel framework, the interior of the cabin was clearly visible, the seats hanging upside down, still bolted to the cabin floor. I shuddered at the thought of whatever fate had befallen the occupants of those seats, some of whom I had probably already seen at the temporary mortuary. I remembered a comment that had been made the evening before: that the badly damaged bodies were those of the lucky ones – they would have been the passengers who died on impact. The intact bodies were more likely the passengers who had survived the impact, only to drown in the submerged wreckage.

On the second night I returned to my room and found a photocopied letter from the hotel management waiting for me. Even though by this time I'd relinquished any sense of reality this letter still took some believing. I really wish I'd kept it and brought it home with me:

"Dear Guest,

Following the recent tragic events and the subsequent wreckage on the beach, the management regret to inform all guests that for safety reasons the hotel watersports department is temporarily closed. All watersports activities and lessons are cancelled until further notice.

Please also take extra care on the beach and if you should find any personal belongings, documents or even body parts then please alert a member of hotel staff immediately.

Thank you."

The letter wasn't the only thing waiting for me. I walked into the bathroom and there, staring back at me, was that same lizard again. I recognised him by a little pale patch down near his tail. I thought that chasing him round with a shoe that first night might've given him a strong hint that he wasn't a welcome room-mate, but his reappearance seemed to demonstrate a disturbing lack of contrition on his behalf. I brandished a towel at him, shouting threats to kick his reptilian little butt, but once again the scaly pimpernel made good his escape.

Each morning I went down to the open air dining area to join the other team members for breakfast. The entire hotel complex was served by a constantly playing music system which, unfortunately, seemed to have only a very small repertoire, played on a constant loop. If I heard Neil Diamond singing "Love On The Rocks" once, I must've heard it a hundred times during my time on the island. However, from a vantage point at the top of the stairs leading down into the alfresco dining area I could see the aircraft tail fin still protruding from the sea and the song title started to feel remarkably appropriate, in a very grim sort of way.

But supermarket-style musack aside, for that precious hour or so each morning we could sit, eat and kid ourselves we were just holidaying guests like all the others who, despite the presence of the wreckage on the beach, clearly weren't going to let the crash spoil *their* two weeks in the sun. I couldn't blame the other guests. After all there was nothing they could do about it anyway. But for me, although breakfast and the evening meal in the beach front dining area were welcome moments of diversion from the horrors that awaited us each day further up the island, the normality of it all still seemed so disrespectful to the men, women and children who had died just yards away from where we sat drinking and eating.

As I have explained, the processing of the bodies consisted of removal from the refrigerated store, listing of gender and any identifying marks, tattoos and clothing, before dental examination began. After the odontologists had compiled a record of dental details, each body, now carrying its number code, would be carried down into the tent erected in the yard for the embalmers to work in, before the newly embalmed body was then securely bagged, carried back into the other half of the warehouse and placed in one of the coffins recently delivered from a South African supplier, whose help the team had enlisted. The coffin would then also be marked with the body number. Although I managed to acclimatise myself to the temperature and humidity, throughout this whole exercise the tropical heat and the smell were still barely tolerable.

I found myself hopelessly out of my depth; although even the veteran team members all said that the working conditions were far in excess of what they might normally expect to find on a disaster response anyway. The Comoros islands were amongst the worst possible places in the world for the crash to have happened. We were stuck on a tiny, backward island many miles from even the continental mainland and the pressure was on to get the work done as quickly and as best we could.

One of the veteran team members had warned me on the outward journey that I must guard against being overcome by the scale of the human tragedy; in other words to guard against becoming overwhelmed by seeing so much death in one go. Ironically though, what hit me most was not the scale of the human tragedy, but quite the opposite - the de-humanising effect I felt the work was having on me. The backwardness of the island and consequent lack of infrastructure and facilities made our working conditions atrocious and it all became little more than an endurance test. I don't mean the deceased were treated with any less dignity – quite the contrary; I just mean that, for me at least, it all became a process with no feeling at all beyond anticipation for getting to the last body. Men, women and children passed through my hands, but by that time they were just bodies, or body parts in some cases. I just wanted it to be over. I had never felt like that about my work before and feeling like that now bothered me far more than the scale of the human tragedy itself.

When I had been faced with an unpleasant body back home there was always the knowledge that I only had to tolerate it for a short time – only for as long as it took to treat and prepare the body and get it bagged and sealed in the coffin. I would grit my teeth, get on with it, put it back in the fridge and get on with something else. But here on the island, there was no end in sight. Every one of the hundred odd bodies was in an appalling state and there weren't even any proper mortuary facilities to make the work more bearable. I vowed that when I returned home I would never again worry about dealing with mutilated, burnt or decomposed cases as it could never again be as dreadful as what I had to cope with on the island.

Mercifully I would soon be diverted away from duties in the embalming tent, instead assigned to examining and listing the clothing and other identifiable marks on bodies, as well as searching and listing the personal documents and passports recovered from the crash site. Later still I was left to organise the coffining of the embalmed bodies. Part of this latter role required me to supervise the three or four local men we had enlisted to act as porters. With typical gallows humour, our temporary porters were nicknamed the "humpers & dumpers." There was a strict rule that they were only to handle bagged or coffined bodies, but such was the poverty on the island they were only too glad to put up with even that work just to earn the small amount of currency we could pay them. We kitted them out in white coats or disposable overalls, rubber gloves and paper masks, all of which seemed rather pointless considering they still then worked with bare feet or flip flops. One of the humpers & dumpers somehow found himself an old fashioned World War Two-style gas mask to wear and whilst I admired his ingenuity, he really did look utterly ridiculous.

The only way I could communicate with our impromptu team of porters was via mangled schoolboy French, augmented with miming actions. I found myself wishing that I'd paid more attention in French lessons at school, as it really was incredibly exhausting trying to make my mangled French understood, let alone all the while miming various actions to go with it. The only consolation was that at least I could let off steam by cursing in my native tongue, safe in the knowledge that if I kept a neutral facial expression then the language barrier would ensure that the humpers & dumpers couldn't understand what I was saying.

One particular task I was given was to be sent to the island's airport on our bus, to locate some additional embalming supplies that were supposed to have been sent over on a flight a few days after we had arrived. The bus driver left me at the tiny airport and promised to return within the hour. The airport was so small and badly administered that I quickly gained unchallenged access to the cargo handling area.

However, my mission to locate the missing supplies proved fruitless because, despite the language barrier, I sensed the cargo handlers weren't being honest with me anyway. I quickly realised that on a poor island like Grand Comoro the cargo handlers must have been well versed in the art of purloining any cargo which looked valuable and saleable. After much searching I gave up and went outside to wait for the bus. I sat myself on the dusty kerb outside the arrivals/departures/deliveries/anyone else/entrance/exit area and surveyed the empty parking lot around me. There weren't that many flights a day anyway and the whole place was deserted. The sight of the huge black volcano looming over the island in the far distance, together with the sun beating down relentlessly upon me, just made me feel even more dejected. I wasn't a confident, seasoned traveller and yet ironically here I was in just about the most remote outpost of humanity anyone could ever expect to find themselves in. At that moment, home, with its comforting normality, seemed further away than ever. I cannot express how mightily relieved I was when the bus finally re-appeared to take me back to the compound.

Feeling embarrassed at returning to the temporary mortuary empty-handed, I reported to the team leader. Fortunately he wasn't too surprised by my tale of failure, already suspicious that the supplies had been purloined at the airport anyway. The missing consignment was eventually found – unsurprisingly at the island's hospital. We got the much-needed embalming chemicals back, but the additional surgical instruments that the embalmers had requested were lost for good, no doubt hidden away and kept as a lucky windfall for the under-equipped hospital.

One strange but actually rather poignant episode involved our little team of humpers & dumpers. They had been tasked with starting a bonfire at the far end of the compound's yard, to burn all the discarded cardboard boxes our embalming supplies had arrived in, together with clinical waste from the embalming tent – including soiled clothing removed from bodies. I was in the coffin storage area, sealing down a coffin lid, when all of a sudden the hot, sticky air reverberated with the sound of loud bangs, like gunfire. There was shouting and minor panic and even the embalmers had rushed out of their tent, holding their bloodied, gloved hands out in front of them as if they had come out in surrender.

Knowing the island's chequered history, we honestly did wonder whether it was gunfire, but the panic quickly subsided when, following another loud bang, the sound was tracked down to the other end of the yard, where one of the team spotted yellow lifejackets, previously removed from bodies, piled on the bonfire. The bangs were coming from un-discharged compressed gas canisters in the discarded lifejackets exploding in the fire.

Whilst there on the island I only saw two of the victims' relatives. One was an Indian gentleman who would sit under the shade of a tree by the compound all day, every day, in the blistering heat, clutching information about his missing relative, waiting in the hope that someone would come out and give him the news that his loved one had been identified.

We also had a young African woman visit the compound to see the bodies of her husband and two children. There was some debate as to how we should conduct the viewing and whether we should try to set up a private area for her. Such an arrangement just wasn't practically possible anyway so, quite rightly in my view, she was simply led into the warehouse where all the sealed coffins were lined up. Just as I suspected would happen she was oblivious to everything around her, save for the three open coffins halfway up one row containing the bodies of her husband and children. She was very grateful for the opportunity to see their bodies before they were repatriated and I firmly believe, assuming she even noticed or gave a damn, that the sight of all the other coffins, far from causing her further distress, might in fact have given her a sense of the enormity and the reality of the crash in which her family had died; that her own personal loss was part of an equally enormous tragedy and that consequently she was not alone in her grief.

There was however one moment of pure comedy during those dreadful days on the island. It happened back at the hotel on the third evening of our deployment. I was the last one down to the dining table, having been delayed by what was to become a nightly battle of wills with my reptilian nemesis and now also some of his equally scaly and nimble fellow gang members. The rest of the team were all sat in a little dining area set in the middle of the swimming pool and reached by a small walkway on its far side. A civil contingencies consultant on the team had flown out after the rest of us and caught up with us at the hotel. Seeing us across the crowded dining area he stuck his hand up in greeting and walked towards us, threading his way through the other diners' tables. Suddenly he disappeared from view with a huge splash. In amongst all the other tables he hadn't seen that our table was in a dining area on a platform in the middle of the pool and as he headed towards us he plunged straight into the water.

Once all the bodies and remains had been processed, embalmed and coffined, they were transported back to the island airport at Moroni and then flown to Addis Abiba – the Ethiopian capital. Getting the coffins off the Comoros Islands was our first priority as the airline's insurers were naturally anxious to have us finished as soon as possible. They wanted all the coffins off the island and back in their own warehouse facilities, from where there were at least some proper facilities and telecommunications to co-ordinate the onward repatriation of all the bodies to their countries of origin.

Our homeward journey required us to make an overnight stop in Addis Abiba, with all the victims' coffins having arrived at Addis Abiba on the flight before ours. Whilst still at the airport in Addis, we were asked to open one of the coffins to make some check or other, before making our way to a local hotel for the night. The relevant coffin was opened and the check duly performed in the airport cargo area's tiny mortuary - a room intended for individual coffins awaiting routine repatriation to be stored in pending their flights. Whilst some of the team members crammed into the tiny mortuary to deal with the coffin in question, the rest of us were stood outside. Whilst we waited I caught my last sight of all the other coffins on a little train of luggage trailers, four coffins to a trailer, still out on the tarmac of the cargo area.

That final, brief moment was a poignant one. I was now in a calmer frame of mind, hugely relieved that my grim expedition was nearly over and that I would soon be back home. In that moment I was finally able to regard those coffins as evidence of the human tragedy which had occurred. The victims were all individual people who had tragically lost their lives in the crash and weren't just human remains that had to be numbered and processed. It felt good to know that I had not lost my sense of humanity after all – it had just been subconsciously suspended to help me survive the whole experience. The bodies of those who had lost their lives were finally on their way home, like I was.

I had achieved another ambition. Had it not been for that opportunity with the team I might never have experienced disaster work – after all, disasters are mercifully rare events. But, in the spirit of being careful what you wish for, achieving this ambition served to show me this was one professional avenue that I was not cut out for – at least in the sense of volunteering to go out with the regular response teams.

I also lost my enthusiasm for being an embalmer during that time, although looking back I regard that as being the hand of fate at work. Within a few years a new business opportunity would mean I no longer had the time to carry out embalming myself anyway, instead having to rely on employing a freelance embalmer. So I suppose the experience with the air crash just made it easier for me to put aside any further ambitions I had as an embalmer and ensured that I would instead concentrate solely on funeral directing duties. There are plenty of good embalmers out there, but in my case my true vocation was to be found outside the embalming room.

But going back to disaster work itself, you must never say never and I would not be remotely surprised if those crazy days spent at that crash are not my last experience with mass fatality work. I learnt a huge amount from my involvement with that incident and that experience has come in useful in all kinds of ways in the years since. But if, or when, there is a next time, it will be because some cataclysm has happened much closer to home and like any other local funeral director, I will respond in whatever manner is required - after all, it's not as if I don't have valuable experience to offer. It's just that I won't go volunteering again. If I'm needed, they can come and find me next time.

CHAPTER FOURTEEN

Home, James

"The lead car is unique, except for the one behind it which is identical."

Murray Walker

One immediate benefit from my expedition with the disaster team was of course that we were paid for our time out on service and this in turn helped towards the purchase of a second-hand limousine. I was hiring a hearse from Fred Stevens Funeral Directors in Nailsworth, with whom I had an agreement for the supply of coffins and bearers, etc. and we agreed that it would be to both our advantages if I in turn ran a limousine. Having purchased the vehicle, a Ford Granada-based conversion, I also offered the vehicle to hire to other funeral directors in the county. Within a short space of time I built up a valuable and useful sideline in limousine hire and over the next three years I served a regular client base of fourteen funeral companies operating around the county, from the furthest town west: Cinderford in the Forest of Dean, across to Cirencester, "Capital of the Cotswolds," in the east, together with Cheltenham in the north and the market town of Dursley in the south.

The Stroud district is largely middle of the road where funerals are concerned, with no real extremes or particular ethnic groups, although its concentrations of artistic and free-thinking communities mean we see far more alternative funerals than funeral directors in other parts of the county might. However, taken as a whole, those three years of limousine hire would fill in some gaps in my experience by taking me on funerals for the military, gypsies/travellers, showmen, the West Indian and Afro-Carribbean communities, together with a broad slice of the social spectrum in both urban and rural areas.

The style of a funeral company is usually a product of the area that it serves, although there were some notable exceptions to that rule! I very quickly got to know the quirks and traditions of each company, ranging from an approach so relaxed that I usually ended up doing half the funeral director's job for him, through to one company whose approach was so militaristic that their funerals were often rather self conscious affairs, with their staff expected to follow a little choreographed routine, especially when in church, of bowing to the altar, bowing to the coffin and generally making slightly too much of a show. The funeral director meanwhile was over-solicitous to the point of appearing patronising and I vividly remember one funeral where, as the limousine driver, I was expected to stand to attention beside my vehicle, laden as it was with the family mourners, whilst the funeral director marched back down the garden path and made a grand show of checking that the grieving family had locked all the doors and windows of their house. However, as I said, those two firms were the extremes and in between were twelve other firms who just quietly got on with doing their job discreetly and well.

Running a new funeral business is a very lonely task, particularly in the very early days when funerals can be few and far between and I missed meeting up with other undertakers for a chat and a gossip at the principal gathering points such as the crematorium or the hospital mortuaries, where I would constantly bump into staff from other firms in my part of the county. Because of this, one of my favourite limousine hire client firms was the large company in Gloucester that Rick, my ex-colleague had originally started his career with. Their team of driver/bearers comprised mostly of young men of around my age and I enjoyed the camaraderie in the staff room in between funerals, as I was often hired for more than one funeral a day with that company, working from either their main office in the city or circulating to any of their four other branches around the county.

Whilst for Stroud funeral directors the question would normally be whether the family even wanted a limousine, for this Gloucester funeral company in particular it was more usually a case of *how many* limousines the family required. One of my first hire jobs with this firm was a West Indian funeral. The church was full, but as I would learn, with West Indian funerals there would usually be just as many people gathered outside the church – all dressed smartly in traditional black, but simply hanging around chatting and smoking.

I was told that the company's funeral arrangers would always word the newspaper death notice with the time of the service half an hour earlier than it actually was, simply because the mourners always seemed to follow their timescale rather than the formally arranged schedule. West Indian and Afro-Carribbean funerals were always very large and sumptuous affairs, with the families always wanting the most expensive items, to the extent that, like gyspy and traveller funerals, there would often be the incongruity of an authentic Victorian horse-drawn hearse transporting a very un-Victorian American-style casket, along with very modern and elaborate floral arrangements.

Another tradition with West Indian and Afro-Carribbean funerals was for the menfolk to backfill the grave themselves whilst the women sang hymns and songs. I've watched as expensive suit jackets have been discarded on the ground whilst the wearer rolled up his sleeves and picked up a shovel. I remember one elderly gentleman, resplendent in sharp suit, waistcoat, hat and even spats, climbing up onto the soil pile, shovelling away with gusto, oblivious to all the muck he was getting on his fancy clothing. The city cemetery department would gratefully leave shovels laid out ready for the mourners to use, whilst staff at the adjacent crematorium would grit their teeth as they watched the main driveway and then all the access roads descend into gridlocked chaos as the parking for the ethnic burials would go on to affect the traffic for the crematorium funerals too. Many is the time when the crematorium chapel attendant has said to me on my own funerals to make sure I make a prompt departure or I'll get hemmed in by the West Indian burial due for arrival at the cemetery.

Whilst this particular company in Gloucester was favoured by the West Indian and Afro-Caribbean families, another funeral company just a few yards down the same city street was favoured by the showmen and to some extent the gypsy and traveller community too. By "showmen" I mean the men and their families who own and operate travelling fairground attractions and rides. There's a showmen's ground tucked under a road bridge down by one of the old riverside suburbs on the western edge of Gloucester and a hire job took me down there one day, along with the hearse and four other limousines. The showmen's ground was their winter base, before the fair season took them travelling around the country with their rides and attractions. It was rather strange to drive round their compound, looking at the dismantled fairground rides sat cheek by jowl with huge and luxurious caravans and the owners' equally expensive cars.

Once again the deceased was in a top of the range solid oak casket surrounded by elaborate flowers and as we drove round the access road in the compound to the family's caravan, I could already see a great volume of flowers being loaded up onto a flat bed lorry by burly young men in smart suits. Once the cortege had set off for the church up in the city, groups of these same young men would drive ahead of us in pick-up trucks, pulling up on verges next to roundabouts and junctions and stopping the traffic for us to pass through unhindered. I can't imagine the police letting my staff do that but unsurprisingly these lads seemed to get away with it. The scene at the church and later at the cemetery, where there was an area set aside for the showmen's graves, was chaotic. The church was full to overflowing, with showmen's families coming from all over the country and there was the same chaos and gridlock around the cemetery and crematorium grounds afterwards.

The term "gypsy" more specifically refers to people of Romani origin, an ethnic group that can be traced to the Indian subcontinent, whilst "travellers" are mainly of Irish or English origin, but from a funeral point of view, the distinctions are largely irrelevant – as long as you don't actually say that to either a gypsy or a traveller of course…

I've done hire jobs with funeral directors acting for both communities and in each case the cortege, consisting of at least five limousines and often a horse-drawn hearse - when distance would allow - would start at the caravan site. It is very hard not to feel intimidated on these caravan sites, but in fairness at a personal level I never actually experienced any problems. Indeed, from the safety of my position as a member of funeral staff, it was grimly fascinating to see these tumultuous communities up close.

On one particular funeral the coffin firstly had to be removed from the deceased's caravan. We had to stay outside (being non-travellers, the caravan was strictly out of bounds to us) and wait for the family members to pass the coffin out through the window – bearing in mind we are talking about a very large caravan to start with! The funeral itself was the usual combination of ostentation, chaos and crowds and when the burial itself was finally completed, myself and the other limousine drivers delivered our passengers to the hotel where the wake was being held. The hotel was accessed by a long driveway and whilst our passengers were getting out of the limousines, the verges of the drive way were starting to fill with overflow parking (there's *always* an overflow of parking at gypsy/traveller funerals). The main car park was already full and there was no room to turn the limousines, leaving us with no choice but to reverse back down the driveway to get out.

I knew that with further carloads of mourners arriving, every second counted if my limousine was to stand a chance of getting out again. The reverse was very tight, with bends in the drive and parked cars either side and whilst the drivers of the third and fourth limousines managed to get out, the driver of the second car was struggling. To be honest I think he would have struggled with a straight line reverse in an empty car park; but as I was the lead car and consequently now the last one to reverse out, I had to grit my teeth and wait for him to extricate himself, all the while having visions of being stuck in that hotel car park all night if the second limo driver didn't get his act together.

My limousine hire work took me right round Gloucestershire, where every part of the county has its own distinct character. A funeral in the south of the county saw the cortege leave the church and make its way up the M5 motorway en route to Gloucester Crematorium. The funeral was running a little tight for time and the funeral director decided that rather than use the A38 trunk road we would instead make up the time by taking the cortege up the M5. We could have retained the dignity of the moment if we had just used other traffic as a pacemaker, but with the hearse and lead limousine weaving across the three lanes and overtaking slower traffic I had no choice but to do the same. The looks on some drivers' faces when they were overtaken by an entire funeral procession was something to behold!

The Cooper's Hill Cheese-Rolling Contest, a tradition believed to be at least 200 years old, is an annual event held on the Spring Bank Holiday at Coopers Hill, between Gloucester and Cheltenham. From the top of the 1:3 gradient hill, a round of Double Gloucester, a hard cheese typically manufactured in cylindrical blocks, is rolled and competitors race down the hill after it. The first person over the finish line at the bottom of the hill wins the cheese. In theory at least, competitors are aiming to catch the cheese, although it can reach speeds up to 70 mph, enough to knock over and injure an unwary spectator. Indeed, each year the event yields an average of twenty casualties – mostly fallen competitors but sometimes spectators injured in collisions too.

Coopers Hill commands stunning views across to the regency town of Cheltenham and it is perhaps a blessing in disguise that I can recall no especially memorable funerals in Cheltenham during the course of my limousine hire work, although one funeral I recall working on there saw me driving the last limousine in a cortege which had stopped in a petrol station whilst the funeral director went into the shop to ask for directions to the area where the bereaved family's house was. Very embarrassing to witness and very unprofessional.

When you're a hire driver you have to put up with the way that your hirer conducts their business but it wasn't until I started carrying out limousine hire work that I had any real insight into what life in other funeral firms was like. It was during that time I realised just how very fortunate I was to have been trained at Thomas Broad & Son. I'm not saying things never went wrong at Broad's – of course they did occasionally, but more often than not it was due to circumstances beyond our control – a tyre puncture on a hearse perhaps, or double bookings at the crematorium, to mention two examples. But even when it was our fault it was invariably due to genuine human error, which sadly we are all liable to make at times. The key difference at Thomas Broad & Son was that we never made any mistakes out of ineptitude or incompetence. The incident with the petrol station would never, ever have happened under Paul and Michael – they wouldn't even contemplate going out on a funeral without knowing exactly where they were going. Routes would be planned and timings worked out. Likewise they knew their churches well too and always planned for any practical issues beforehand.

But the biggest difference of all was the mindset at Thomas Broad & Son. Paul and Michael, much to my dismay, may have been reluctant to embrace some of the finishing touches that families really do appreciate, like walking in front of the hearse for example, or if there were large amounts of flowers, then displaying them on the roof of the hearse rather than having the excess flowers taken to the graveside in a separate vehice before the funeral. But nevertheless, for all that other firms would be good at putting on something of a show, there were some that would then let themselves down with the basics. Paul and Michael would never allow the cortege to go crawling down a street looking for the right house number for example; they would've found the house beforehand. Likewise they would always plan journeys to ensure that we knew exactly where we were going, that we arrived on time at each location and that we would be on time at the church or crematorium.

I was ingrained with the attitude that the basics mattered more than showmanship. Gleaming fleets of expensive cars, staff in nice corporate uniforms, elaborate coffins, they all count for nothing if the undertakers can't even manage to find the right house or ensure that the cortege arrives on time at the church. Paul and Michael's style might have been low-key, but they more than made up for it by bringing common sense, experience and most importantly, emotional intelligence, to their work. These were all qualities which I took for granted were to be learnt before I could conduct funerals and it was only after I'd left Broad's that I began to realise just how fortunate I was to have been ingrained with those standards.

I'd been given a very solid foundation upon which to build my own approach, but because a funeral is a ritual, it does still need a sense of style and occasion. So, being anxious to move away from the sensible but utilitarian approach that I had been trained under and wanting to bring some style back into the equation, I quickly realised that the key was to judge the moment, identify what tone the occasion had, what style would suit it best and then ensuring that my planning and presentation would still allow the ritual to unfold in a way that made it the deceased's and the family's show, not mine.

A part of Gloucestershire I always enjoyed doing hire work in was the slightly magical and other-worldly Forest Of Dean, sitting high on its plateau across the river Severn, in the west of the county. One of England's few remaining ancient forests, it enjoys a spectacular natural beauty and has been an inspiration to many writers including JRR Tolkien, Dennis Potter and JK Rowling.

The Forest is defined by its distinctive pine tree-topped hills and drab, grey-looking, slate roofed villages and towns so indicative of its proximity to Wales. Indeed this borderland, valued for its resources of iron, coal, stone and timber was fought over for centuries by the English and the Welsh.

The forest is bounded by the River Wye to its western flank and the Severn of course to its east, so historically the area was considered isolated and self-contained, with a very parochial reputation. That has all changed now, but the Forest still retains a distinct feel and doing hire work down there always felt like a real expedition. On one funeral in a cramped Non-Conformist Chapel in Cinderford, the coffin became firmly wedged in the narrow vestibule doorway and it took a few tense moments and crushed fingers to edge the coffin through the door an inch at a time, with just two of us bearers taking an end each, before we finally managed to get the deceased up the aisle for his funeral.

In stark comparison were the chocolate box villages and towns built of honey coloured stone up in the Cotswold Hills in the north and east of Gloucestershire. This is international tourist country and a double funeral (an elderly lady and her adult son who lost their lives in a house fire), held at the main parish church in Cirencester town centre, became the focus of much attention from a delighted crowd of video camera-wielding Japanese tourists who happened to be outside the church when myself and the other bearers shouldered the coffins out to the waiting hearses.

Hire jobs for Leonard Parry, the elderly funeral director in Tetbury, the Cotswold town made famous because of nearby Highgrove House, the country home of HRH Prince Charles, always managed to provide me with some light relief from the intensity of the urban funerals. Mr. Parry undoubtedly had the most picturesque trading area - and by far the most gentrified – and it was always a pleasure to work on funerals in Tetbury or its surrounding areas. It always amused me that the local townsfolk would question why Mr. Parry didn't just retire, remarking on his age and associated deafness, yet it still seemed an article of faith that when need arose those same families would still use him in favour of his younger, more alert competitors.

On one of his funerals, we were at Beverstone, a tiny but perfect Cotswold village, completely untouched save for the main road that cuts it neatly in half. The small village church is overlooked by a castle - originally a medieval stone fortress. Beverstone Castle always fascinated me; its presence largely invisible from the road but clearly viewed from the church. Much of it is in a state of ruin, although a portion of the structure is still occupied. The ruins loom over the church and if ever there was a place that felt like it should be heavily haunted then the castle, the church and their surroundings had to be it.

At the close of the funeral service the vicar intoned the words of the Nunc Dimittis:

"Now, Lord, let your servant go in peace: your word has been fulfilled. My own eyes have seen the salvation which you have prepared in the sight of every people; a light to reveal you to the nations and the glory of your people Israel. Glory to the Father and to the Son and to the Holy Spirit; as it was in the beginning is now and shall be for ever. Amen."

Mr. Parry, myself and the rest of the bearers were stood just out of sight of the priest and there was a pregnant pause, which I knew was the moment for us to step forward and turn the coffin, ready to carry it from the church. But elderly Mr. Parry hadn't heard the priest. I knew from past experience that the aged funeral director didn't take kindly to any of his staff taking the initiative so I stood stock still with the other bearers. In a tone of voice that implied he didn't want to spoil the moment, the priest said,

"The funeral director will now lead the coffin from the church."

Silence and stillness. Leonard Parry didn't like anyone having the temerity to even open the tailgate of the hearse, as that was *his* job, so I thought, *no, I'm staying put and waiting for instructions.*

"THE FUNERAL DIRECTOR WILL NOW LEAD THE COFFIN FROM THE CHURCH."

Still the old man didn't catch it. The mourners were all staring at us and I could hear the nervous shuffle of feet. A mourner made a signal at us and then old Mr. Parry cottoned on and gave us the signal to move forward.

"IS THE FUNERAL DIRECT', AH! THANK YOU. We will all now walk in procession from the church."

Back in the much less inspiring surroundings of a suburban housing estate in Gloucester, I was on hire with my limousine for the funeral of a teenage lad who had died in a road accident, his car having collided with a refuse collection lorry. Mine was the only limousine on the funeral so I had the immediate family – the lad's mother and sisters - in the car with me. Naturally, they were distressed enough as it was, but as we made our way out of their estate, the cortege became stuck behind, of all things, a refuse collection in progress. Because of parked cars there was no way round the dustbin lorry and so I had to grit my teeth and put up with it until we reached the junction at the top of the road. The mother was understandably becoming increasingly angry and distressed but there was little I could do or say. Unfortunately, as if fate hadn't already dealt her a dreadful hand with the loss of her son, good fortune would continue to be conspicuous by its absence throughout the funeral.

During the funeral service itself a recording of one of her son's favourite songs jammed in the CD player and the techno-phobic church verger, poor woman, had a minor panic as she tried to restart the machine. At the cemetery, by which time of course, the tension and distress in the air was palpable anyway, the minister began the prayers of committal before pausing whilst we lowered the coffin. At this particular cemetery – flat in a way that those of us who bury in the hill country of Stroud could only dream of - the graves were all dug with a mechanical digger, so even the single-depth graves were always generously deep.

I was at the foot end of the coffin and as we lowered it into the grave, my end of the lowering strap suddenly went slack in my hand and there was an almighty thump as the foot end of the coffin suddenly dropped and hit the bottom of the grave. I looked down and saw part of the decorative wooden moulding surrounding the coffin lid had split and sprung free. The combination of the noise and the visible damage to the coffin made the whole moment a very ugly one. There had been an awful gasp from the lad's mother as her son's coffin dropped the last ten inches to the bottom of the grave. It was a dreadful moment but it was too late to do anything.

As soon as the minister finished the prayers and we bearers had stepped away from the grave, myself and the other bearer on the foot end with me ducked away out of earshot where he quickly apologised and explained that because the grave was so deep he had run out of lowering strap. He could feel the end of the strap go through his hand but just couldn't get enough grip to hold it with his fingers and so the coffin dropped, with me left on the other side of the grave holding a now redundant strap. There is never a time that you want something like that to happen, but on that funeral of all funerals! These accidents do happen occasionally – that same thing happened to one of my bearers on a funeral I was conducting once – and sadly there is nothing you can do to rewind the fateful moment.

One funeral where pent-up emotions manifested themselves in a totally unexpected manner was that of a child who had died following a road accident. The funeral itself was, as might be imagined, attended by hundreds of mourners. The family-owned company I was on hire to already had enough of their own bearers on hand, so I stayed outside the church and busied myself turning the vehicles ready for departure to the local cemetery. It was very sad and touching to see all the different floral tributes in the hearse, including the obligatory teddy bear designs and even a floral tribute made to represent the "Teletubbies' Garden" from the well known childrens' television show, complete with little figures of the various Teletubby characters.

Once we had made our way to the cemetery and the small white coffin had been lowered into the grave, myself and the bearers placed all the flowers out on the grass for the family to see, before we retreated back to a respectful distance. I was aware that there was some kind of family controversy but I didn't know any of the details. However, as the child's parents made their way down the line of flowers, the father suddenly became very irate upon reading the card on one of the floral tributes. He started shouting and swearing but from where I was stood over by the limousine all I could make out was that he was extremely angry that the flowers were sent by someone to whom he clearly bore some animosity.

To everyone's utter amazement the father picked up the offending floral tribute, held it out in front of him and drop-kicked it across the cemetery. The flowers exploded mid-air in a burst of petals and greenery and the shocked mourners were left open-jawed and speechless. From where I was standing the whole moment seemed to happen in slow motion.

I had encountered similar fractious situations before, often caused by second marriages and the resultant ex-partners of the deceased, where families have told me that "if he/she sends any flowers you're not to bring them anywhere near the funeral. Just dump them." Easy for them to say, of course. But actually kicking the flowers across the cemetery...? Well, that was a new one for me. My reaction quickly switched from embarrassed horror to relief though, as I reminded myself that at least I was driving the second limousine – because for the leading limousine driver it was going to be a very tense journey back to the house, having to transport the bereaved father in that state of emotion.

Driving a limousine on funerals almost invariably means having to act as a bearer as well. One funeral, for a young woman of very generous physical proportions, became memorable for a variety of reasons. We carried her coffin into church to the sound of Gloria Gaynor singing "I Will Survive" and whilst I appreciated that the deceased was something of a disco diva and that was her favorite song, the choice of music was more than a little jarring and I could see some of the mourners squirming with embarrassment. The firm I was on hire to were notorious for only ever putting four bearers on funerals regardless of the size of the coffin. Any sensible funeral director would have increased the bearers to six in the case of extremely heavy coffins like this one was, particularly as the path leading up to the church was so long as well.

Anyway, we struggled in to the dulcet tones of Ms. Gaynor and with enormous relief placed the coffin on the trestles. The church was absolutely full. We retreated outside until the close of the service, before re-entering the church, turning the coffin and lifting it onto our shoulders. Somehow or other, instead of the vicar leading everyone out of church as normal, the congregation somehow misunderstood his instructions and began filing out of church from the back rows first, instead of following the family.

The vicar and the funeral director decided that there was no point in trying to stop everyone, leaving us with the weighty coffin still on our shoulders and the family stuck in their pews whilst the rest of the congregation filed out. The coffin was becoming unbearably heavy and I couldn't hold it on my shoulders any longer. I hissed to the funeral director to put the trestles back under it as I started to move the coffin off my shoulder, hoping that the other bearers would realise what I was trying to do.

With all the mourners then gathered outside, the embarrassed vicar asked the family to follow out behind the coffin. As we went to lift the coffin back onto our shoulders I found to my horror that I had lost all feeling and strength in my arm and it was only with some ungainly struggling and another hiss for help to the funeral director that I was able to get the coffin back up onto my shoulder.

As we moved off I could hear a knocking noise from inside the coffin. It wouldn't stop and as we got to the graveside it grew louder and more persistent and I could see that even some of the mourners were wondering what on earth it was. As we placed the coffin on the trestles by the grave ready to put the lowering straps round it I muttered an enquiry about the noise to one of the other bearers. He whispered back that it was a ball inside the coffin. Apparently it was placed in the coffin by the family as a memento when they visited the chapel of rest. The bearer whispered further, in none too elegant terms, that the body filled the coffin so well that the ball had nowhere to sit and was instead rolling around on top of the body. I never did get to ask the funeral director why the family saw fit to place a ball in the coffin, but then I have seen so many different items placed in coffins over the years that I can only assume this young woman's family must have had their reasons too.

Wootton Bassett: a small, picturesque, but otherwise unremarkable Wiltshire market town that will now forever be defined by the military repatriation processions along its High Street. The town, later renamed Royal Wotton Bassett in honour of its role in bringing the UK's military dead back home, has a bit part in the story of one very memorable limousine hire job that I carried out in the late 1990's, long before Wootton Bassett came to national prominence with the repatriations.

At the beginning of this chapter I mentioned a funeral company with a very militaristic approach to conducting funerals and considering the funeral they had hired me for was to be held with full military honours, you would have thought that they would be well suited for the occasion. I beg to differ...

When I was studying for the Diploma In Funeral Directing the subject of military funerals was, obviously, considered a matter for special consideration. I always remember my tutor telling us that rule number one for conducting a military funeral is that the funeral director doesn't conduct a military funeral. The military do. The point being that beyond the normal preparation of the body and the coffin, the military don't need civilian funeral directors interfering. If they can move a thousand men, machines, tanks and equipment to any environment in the world and fight a war, or if they can execute state ceremonials like Trooping The Colour with total precision, then they certainly don't need a civilian funeral director to tell them how to conduct a military funeral. This had always struck me as blindingly obvious, but, as I would discover, not everyone gets the point.

The deceased was stationed at RAF Lyneham, just down the road from Wootton Bassett of course, although he actually lived in a Gloucestershire town served by the funeral company in question. He was a member of the ground crew at Lyneham and was a long serving and highly respected member of the team at the base. His sudden death from natural causes, whilst still in service, entitled him to a military funeral. Mine was to be one of four limousines and my vehicle was appointed as the lead car, taking the immediate family – the wife and children. It was to be an all day engagement, but I was aware and prepared for that.

The four of us drivers gathered in the funeral home garage where we were given a briefing on the funeral. With the help of a whiteboard, Edward, the funeral director, drew a diagram of the parking arrangements for the hearse and limousines outside the church. I remember hoping that Edward was simply repeating what the RAF had told him they wanted us to do and not just setting out his own plan of how *he* thought the funeral should be approached.

The cortege set off to the house and everything went smoothly there. With the six family members sat in the passenger compartment of my limousine I also had a young RAF officer, acting as a family liaison, sat in the front with me. Most of the journey down into Wiltshire was spent with him and the family sharing reminiscences of the deceased, but I thought the slightly precious and very plummy-sounding officer was being a little too gushing about the deceased and I wondered how well he actually knew him.

The funeral service was held at a church adjacent to the RAF base and on arrival the parking plan began to make sense as the military police had temporarily closed the main road to allow us to park, disembark our passengers and wait whilst the RAF bearer party carried the coffin from the hearse into the church. As soon as the mourners were sufficiently far up the church path we whisked the vehicles away to allow the road to be re-opened. Following the service, the military police held up the traffic again as the coffin was placed in the hearse and with the family gathered up we set off back to the deceased's home town for the burial.

The route back, as it was on the way down, took us through Wootton Bassett, along that picturesque and now famous High Street. We attracted much attention coming away from the base, what with the military police stopping the traffic outside the base and then the sight of a union flag bedecked coffin in the back of the hearse – bearing in mind that back then, in the late nineties, such sights were synonymous not with military repatriations but reminiscent of the return home of The Late Princess Diana from Paris just twelve months earlier.

In fairness, Edward the funeral director had behaved himself so far. I noticed him twitching and itching with eagerness when we arrived at the church but everything was so tightly planned that there was little chance for him to interfere with anything, as the military bearer party and commanding officers were of course already lined up by the roadside waiting in readiness.

The burial itself was quite interesting. With a military bearer party every little movement is carried out as a specific manoeuvre, even down to the bearers shuffling sideways to position themselves perfectly by the hearse ready to withdraw the coffin. Taking the coffin across to the grave and placing it on the grave ready to be lowered -which would normally take my bearers an unhurried and dignified three or four minutes - took nearer ten or twelve minutes in military style. The ceremonial was very impressive, but by this stage of a funeral every second will feel like an eternity to the immediate family and I was especially conscious on that particular occasion that the deceased's wife was starting to buckle under the emotion of those excruciatingly long moments.

Finally the burial was completed and the last post was played by the military bugler. Then came the bit I was looking forward to and about which Edward had told us during our pre-funeral briefing: The RAF had arranged for a Hercules aircraft to make a tribute fly-past over the cemetery.

The sound of that huge aircraft as it roared overhead, perfectly timed of course, sent a shiver of excitement down my spine and it was indeed a fine tribute. Then it was back to the limousine and the journey back down to RAF Lyneham again for the family to rejoin all the other mourners for refreshments in the Officer's Mess. Throughout the day's journeys, from the house down to Lyneham, back up into Gloucestershire and now back down to Lyneham again, the liaison officer and I had barely exchanged a word, which was fine by me, but on the journey back to Lyneham I once again had to listen to him playing up to his audience in the back of the car; in particular the deceased's twenty-something daughter who was, to put it kindly, not the brightest star in the sky.

"Weren't that lovely, when that plane came over. Amazin' innit, how a plane was flyin' past when Dad was buried. Was that one from Lyneham?" She asked.

"That was actually a ceremonial fly past." The officer replied tartly. "The Hercules aircraft was attached to one of the squadrons at Lyneham. We couldn't let one of our own go without saying goodbye."

"That was one of 'is?! Oh, wan' that nice of 'em."

We still had a depressingly long way to travel at that point and the combination of slow driving whilst having to keep the other limo's in procession with me, was starting to get tiring. Meanwhile, the conversation stumbled on.

"Blimey! Weren't it big though!" The daughter said, sounding impressed.

"Well yes, it is our primary cargo carrying aircraft, designed for use in all theatres of conflict."

"Ohhh, riiight. So 'oo was in it then, was it just the driver?"

"The Hercules actually has a crew of three, including the, ahem, *pilot* rather than driver and the co-pilot and loadmaster." The liaison officer explained rather pointedly, clearly now tiring of the daughter's inane questions.

A military police car was waiting at Lyneham's main gate to lead us through the base to the Officers' Mess and after we pulled up I held open the limo door for the family. Meanwhile Edward the funeral director, who had been riding in the second limousine, made a move for the Mess door, doubtless to hold it open for the family, but he was instantly and not entirely discreetly halted by another RAF officer who sharply thrust his ceremonial baton across the eager funeral director's chest and stopped him dead in his tracks. The look on Edward's face was priceless.

Rule number one of military funerals Edward, I thought, *rule number one.*

After a fairly pleasant bit of down time in another part of the Mess, where Edward, myself and the other limousine drivers had been left on our own with our packed lunches, we were eventually given the request to take the family home again. It was a very long day, fascinating at times and quietly hilarious at others! Best of all, the liaison officer stayed behind at the base when I took the family on the final journey home. With the family now out of earshot behind the firmly closed glass partition I could at least enjoy the Wiltshire scenery in peace, taking in the picturesque street scene of Wootton Bassett High Street one last time. I'm sure the hearse drivers who have passed up that High Street so many times since will have their own stories to tell.

CHAPTER FIFTEEN

Millennium Bug

"People think responsibility is hard to bear. It's not. I think that sometimes it is the absence of responsibility that is harder to bear. You have a great feeling of impotence."

Henry Kissinger – American Political Scientist

There was nothing remotely unusual about that Thursday morning in June 1999 until the front door of my funeral home swung open and William Stevens, the funeral director from Nailsworth, walked in. He hovered hesitantly in the reception area for a moment, greeted my mother, who was sat behind the reception desk and made small talk before asking if he could speak to me in the private office. I closed the door behind us and we sat down in the comfy seats around the small coffee table.

Without any preamble William announced that he was planning to retire at the end of the year and he was offering me first refusal to buy his family business, Fred Stevens Funeral Directors. I was momentarily stunned as the enormity of what he was saying hit me. I immediately said yes, of course I would want to purchase the company. William broadly outlined the proposed basics: a 1st December take-over date, allowing a hand-over period lasting for the duration of that month, with his official retirement beginning on 31st December. William went on to say he had instructed a commercial valuation company and that he would let me have the sale figure when it was confirmed.

It was a surprisingly brief meeting and he was gone as quickly as he had arrived, leaving me in state of euphoria, my mind already racing with all manner of thoughts. Nailsworth, four miles from Stroud, was very much a separate community and as such Fred Stevens Funeral Directors had its own distinct trading area within the larger Stroud district.

Over the previous decade Nailsworth had been reinvigorated, largely through becoming a food-orientated town, with numerous restaurants and cafes opening, together with many other unusual and high-quality shops. Despite Stroud being the much larger town, it was fast becoming Nailsworth's poor relation as a shopping destination. Meanwhile, having already been working with William at Fred Stevens for the previous seven years, hiring vehicles to each other, along with holiday cover, etc., it all meant I knew the business well. It was quite some prize.

After seven long years of only small amounts of progress it looked as though the breakthrough I had been waiting for had finally arrived, albeit not in quite the way I had imagined. My previous employers, Thomas Broad & Son, were still at the height of their powers and although their virtual monopoly within the district was causing mutterings of dissent in some quarters, I had already learnt the hard way not to underestimate the strength of their reputation.

It took two excruciatingly long months for the valuation on Fred Stevens to finally arrive, by which time I had engaged in many long conversations with my father and in turn with my investors. I naively thought the prospect of being able to purchase Fred Stevens, with its established flow of funerals and the consequent financial security that would bring, would enable the remaining months of 1999 to pass with a spirit of contented expectation – but, as with all things, it didn't quite pan out the way I had imagined.

Rather than peace of mind from knowing that the future suddenly looked brighter, there was in fact nothing but stress and anxiety as this longed for opportunity still hung in the balance until all the financial negotiations were agreed and the legalities completed. Only when we finally got to sign on the dotted line would it all be for real. Until then anything could still happen to scupper the deal. Happily, matters did progress successfully and finally the deal was done. Then we could relax a little and work on more enjoyable matters like planning how we would re-position the company ready for the 21st century.

This was all happening in the run-up to the year 2000 – the new Millennium and there was great anxiety about the so-called Millennium Bug, the name given to the potential problem caused by computer hardware and software which used only the last two digits of the year rather than all four. It was thought that essential computer systems could have been vulnerable when they ticked over from "99" to "00" - making some of them interpret the year as 1900. Pre-Millennium Eve estimates of the likely damage that could have been caused ranged from minor computer irritations to complete global meltdown. In the event there was of course no global meltdown, aircraft did not fall from the sky and the world survived wholly unscathed. I do however remember one pre-takeover meeting with William, when with the potential threat of the Millennium Bug in mind, I was concerned in case the business had any electrical equipment that might be affected. I asked him "If the business has anything that might not be working after midnight on New Year's Eve 1999," before following the question with "...apart from you of course!"

December finally arrived and our move to Nailsworth took place. Organised chaos ensued as we moved into our new home (the adjacent residential property came with the business) and whilst I adjusted to a new business reality, I still had to get on with the normal routine of arranging and conducting funerals. That first week was eventful to say the least. Just days before the move I had been requested to handle the funeral arrangements of a young woman who had died in tragic circumstances, just a month after the death of her mother. Taking on these funeral arrangements put me right in the crossfire of an acrimonious split within her family.

The funeral order, made by the young woman's husband, had come through my Stroud office, Lansdown Funeral Service. It was there that I met the young woman's mother-in-law and sister-in-law one evening so they could dress the young woman's body in her own clothes. They were both nurses and both familiar with the physical reality of death, but few moments were more poignant than the hour the three of us spent restoring the young woman to a more dignified semblance of her former self, before placing her in the coffin.

However, the funeral arrangements were soon to move from the sublime to the ridiculous. It was my first working day after the move to Nailsworth - a dark, stormy December afternoon, when I received an angry and irrational phone call from the partner of the young woman's recently deceased mother, stating that he was aware that the young woman was to be buried next to her mother. He went on to threaten me that if so much as a single flower was disturbed on his partner's (the young woman's mother) new grave there would be hell to pay.

I very politely told him that I was not in the habit of casually desecrating other people's graves. I reiterated that the young woman was to be buried in a new grave adjacent to her mother's and that there was no need for, let alone any intention of, disturbing her mother's grave. I told him that my only concern was to conduct the young woman's funeral in the same dignified manner as any other funeral and that the family politics were no concern of mine. He repeated his threat and I wasted half an hour of my life repeating to him all that I had already said, before firmly explaining that as the funeral director I was not taking anyone's side and that whatever friction existed between him and the rest of the young woman's family was no concern of mine.

I had just moved house, I had just moved office, I was trying to deal with various funeral arrangements, I had telephone engineers getting under my feet installing our new phone system and on top of all that I now had childish nonsense like this to deal with. I was well aware of the rift in the family, but the young woman's husband and the other family members that I had contact with had all been reasonable, co-operative and polite and that was as much as I could possibly ask of any family.

The church was packed for the young woman's funeral and thankfully all went well. My duties were carried out efficiently and sensitively and whether there were further family ructions after that I didn't know, nor was it necessary for me to know. As it was, I had my newly acquired business needing all my attention.

We had moved on the Friday, allowing us the weekend to get over the house move before taking over the office on the Monday morning. On the Thursday before the move William Stevens took a first call from a woman who had lost her partner and with this set to become my first funeral after the take-over, I went with William to perform the removal from the deceased's home.

The funeral itself was a relatively simple set of arrangements, with the funeral taking place entirely at the crematorium. Sadly, the partner had been unable to work due to his illness and his surviving partner was herself in a low paid job. I was never paid in full for that funeral. First irrational threats on the telephone, then my first funeral with my newly acquired business later ended with a bad debt. What an auspicious start! However, the rule that bad things always come in three's was soon to be proved, with a phone call from a lady asking if we could make funeral arrangements. "Yes, of course," I replied.

I took down all the details before the lady then stated that the funeral had to be on a certain day at a certain time. This kind of thing doesn't happen very often, as most families have the sense to know that we as funeral directors need to have at least some involvement in the fixing of dates and times to ensure that we are available ourselves. When these occasions do occur, we can often do a bit of juggling to keep everyone happy, but this time around the date and time the lady specified clashed squarely with another funeral that I had already scheduled.

"Ok, not to worry, we'll try someone else," she said, not giving me a chance to ascertain whether she had even checked whether the vicar and the church were available to fit in with her plan. She had put the cart in front of the horse anyway, but I was never able to find out if in fact we could have helped her after all. It was the icing on a bitter cake that week.

The premises I had purchased with the business were smart and respectable, but very small. They were situated behind an old cloth mill previously owned by the Stevens family. The original chapel of rest had been located in one part of the mill, with a coffin-making workshop housed in another part. Meanwhile the entire funeral business was run alongside a painting and decorating shop in the town, until the shop was closed in 1986 and the decision made to run the funeral side of the business as a full-time concern. Construction of small purpose-built premises followed and it was these premises which I in turn inherited with the business. Meanwhile, the neighbouring mill itself had also been sold, for conversion into the town's new doctors' surgery when the two local medical practices merged.

The funeral premises were built on the basis that all the office administration was done in the adjacent residential property, with arrangement meetings then usually being held at clients' homes. Thus the premises themselves consisted of a small reception room, from which visitors would then enter the chapel of rest and the rest of the building housed the garage, with a curtain at the rear behind which the coffin storage area and a three body fridge were hidden away. My first priority was therefore to establish a small office in the reception room of the premises and reclaim the residential property as a home for my wife Frances and myself.

The fact that the local undertaker was situated immediately behind the local doctors' surgery caused great amusement in Nailsworth – indeed still does to some extent. When giving new visitors directions I would simply pass off this strange anomaly as an "accident of history," before going on to set the record straight by saying that "We were here first." However, it was, and still is, hugely helpful to us of course, although not in the way most people imagine!

For one thing it is very handy for chasing up doctors to get cremation forms signed. However, the only down side comes every year with the advent of the annual flu clinics, when over a period of weeks literally hundreds of patients will be called in for their flu jabs. Our old office door was right opposite one of the surgery's staff entrances, also used for the flu clinic patients to prevent the main reception area being overwhelmed and we often used to get confused, elderly patients coming through our door asking if they had the right place for their flu jab!

Nowadays, the original funeral premises are used as an annexe, housing the garage, coffin workshop and our private mortuary, whilst we also have a brand new two-storey building next door, housing greatly enlarged visitor facilities, new chapels of rest and a new office upstairs, from which we can now watch the chaos down below as the flu clinic patients, many of them elderly drivers, try to negotiate the car park.

By the middle of the first year in Nailsworth I was really feeling the strain. After leaving my employers to set up on my own first time around, my professional metabolism had to adjust to a much slower pace for seven years. But then following the move to Nailsworth I had to pick up the pace again. I continued to have a huge amount of help from my long suffering mother, who looked after the office, together with my wife Frances who helped with answering phones and even the odd removal, whilst my father looked after the accounts and finances – a far from inconsiderable task in itself. However, I was still responsible for arranging and conducting funerals, call-outs round the clock, cleaning and maintaining the vehicles, preparing coffins and all the mortuary work.

So, with the strain showing on me, Frances urged me to go and see my GP, whereupon I was diagnosed as suffering from stress. The doctor advised that I went away for a weekend (about all the time I could spare), away from everything, to get some rest. With Frances and my parents holding the fort I took myself off to my native Wiltshire for the weekend. I think there were concerns about my general state of mind at the time, but I was simply exhausted and just needed a little time to relax and re-balance myself. However, there couldn't have been a worse moment for the bizarre coincidence that was about to take place in my absence.

Frances took a phone call at about half past midnight on the first night I was away. Just like I did at such times of the night, she too answered the phone in still-half-asleep-autopilot-mode. She didn't catch the caller's words of introduction, but she was awake enough to catch the phrase "the name of the deceased is Mr. James Baker." In that instant, she later told me, she suddenly became very much awake! Mercifully it transpired that the caller was one of our local nursing homes and the James Baker in question was a very elderly – and now deceased - gentleman in their care.

It was a very bizarre experience to come home again after my weekend away from it all to see my own name on a card on the fridge door. It was equally strange to have to engrave my name on a coffin nameplate. Mr. Baker was a bachelor and over the course of the next year or so I would go on to bury both of his spinster sisters as well. They were a quaint little trio, living in a very dated little bungalow, perched on one side of a precipitous valley in a village just above our town. When I removed the second sister from the bungalow, the last of the three to die, I knew instantly that the estate agents would be rubbing their hands with glee at the prospect of selling that property, tucked away in its secluded location with stunning views into the valley below.

Indeed, on a later occasion I would find myself at a cottage just a little further back along the side of the valley, sitting on the decking balcony that my bereaved client had built at the foot of his garden. As we sat looking down over that almost vertically-sided valley, the birds flying through the valley were quite literally flying below us.

Those first two years running the newly acquired business were very difficult, as the twenty four hour a day, seven day a week nature of the business left Frances and I with little time to get on with much needed decorating work in the bungalow to make it our home, let alone also trying to work in an office barely large enough for one person. With great relief, during the second year we were finally able to complete modifications to the premises that we had planned to carry out once we'd got settled in to the business. After the work was completed the premises boasted a larger, more sensible space for the office, together with a proper mortuary. Even then, the office still had to double up as a reception area and visitors to the chapel of rest also still had to pass through the office. The significance of that little detail will become clear at the end of this chapter.

The height of that second summer heralded a strange and hectic time, with three very tragic deaths following one after the other, starting with a motorcyclist killed in a road collision, then a paramedic who took his own life and finally an amateur mountaineer who lost his life with another man whilst climbing Mont Blanc in France. All three funerals were of course attended by large numbers of people and each of the three incidents had its own unique challenges for me as a funeral director, but undoubtedly the image that sticks in my mind the most is that of six paramedics in uniform carrying the coffin of their colleague past a guard of honour formed by other paramedics. As one of the churchwardens remarked to me, it seemed so wrong and so poignant to watch professional lifesavers having to carry the coffin of one of their own.

Once I had taken over the business in Nailsworth, I also then inherited the team of part-time bearers. The journey back from Gloucester Crematorium one rainy Thursday afternoon was made all the more grim when a chance remark from one of the elderly bearers as we made our way down the crematorium drive, started a long conversation about gardening implements. From his seat in the rear of the hearse, George had seen one of the crematorium maintenance staff digging an ornamental flower bed and then for the next forty minutes I was forced to listen to George, Len the hearse driver and Roy, the other bearer sat in the rear, debating the merits of everything from spades to lawn edgers. I lost the will to live long before we got back to the office.

Whilst on the subject of the bearers, perhaps it was their age but very often even the simplest of instructions just seemed to go in one ear and straight out of the other. For one funeral we had arrived outside a pictursesque hillside cottage just below the large village of Minchinhampton, on the hill above Nailsworth. As I got out of the hearse in the steep lane outside the cottage, I specifically told the driver, Tony, to wait where he was, to give Clive the limousine driver room to pull up behind him and reverse into the turning for the cottage; after that I would then signal for Tony to reverse the hearse back round as well. No sooner had I shut the hearse door and turned to beckon Clive forward than Tony began reversing the hearse and the two vehicles crunched into each other in the middle of the lane. Fortunately the family didn't hear the collision from inside their cottage and the damage to the vehicles was such that, apart from a loosened bumper on the limousine and a cracked indicator light on the hearse, we had got away without sustaining enough damage to prevent either vehicle carrying on with the funeral.

Within a few years of my move to Nailsworth the rumours began circulating in local funeral profession circles that men in flashy cars with clipboards and laser measuring devices had been seen surveying Dartford House, the premises of my former employers Thomas Broad & Son. The rumours of a take-over became ever more frenetic and then, a phone call one evening, from my freelance embalmer, who also worked for various local branches of a large funeral group, confirmed that Thomas Broad & Son had indeed been sold to this large group.

A year or so later I received a phone call from Paul, my ex-employer, who by then was divorced from Thomas Broad II's daughter Elizabeth. Paul started by informing me that as I was no doubt aware, he was now no longer involved with the old firm or indeed the funeral profession as a whole and so would I be able to handle the funeral arrangements of his second wife's father? Naturally, I said yes, of course, I would be privileged to handle things and a time for me to meet with the family was duly agreed.

I was alone in the office at the time and after I put the phone down I sat silently for a moment, contemplating what had just happened. It was of course a privilege, but also a great responsibility. It was eighteen years previously that I had met Paul for the first time on that fateful first day at his funeral home. As a nervous work experience lad I'd been in awe of everyone and everything and indeed remained so for some time afterwards as I had trained and learnt my way in my new profession. I had listened in fascination as, over those early years, Paul had told me of the various deaths he had attended, of gruesome removals and eventful funerals. Now, after years of hard work, all of those things I too had seen and done. I had come of age, professionally speaking. Now I was the experienced funeral director & business owner and Paul wanted my help.

The meeting with the family was a long one. Paul's father-in-law had left a wife and four children and naturally each wanted to contribute to the planning. In earlier years the meeting would have been a real test, but by now I had the experience and the confidence to referee the meeting and ensure everyone was able to air their views. It took a lot of patience at times but in letting them all talk I was able to get a real sense of what would be needed to make the funeral what they wanted it to be. The various members of the family each had specific ideas and requests, chief amongst which was to reflect their father's involvement with the motor trade by having a stand-out fleet of vehicles, preferably Rolls Royce or Daimlers. They wanted four limousines to follow the hearse and to top it all the funeral was to be at a church in town notorious for its lack of parking.

I left the house with my head spinning just from the verbal bombardment of having so many family members together in one room, but even as I got back in my car I was mentally drawing up a preliminary list of must-do's. First thing the following morning I began making phone calls to assemble a distinctive fleet of vehicles. Having tried my first choice of carriage master without success, I accepted I wasn't going to be able to achieve a wholly matching fleet, so I began a bit of lateral thinking. After looking through trade journal advertisements I hired a vintage Rolls Royce hearse from a funeral company in Hertfordshire and it was through them that I found another hire company who were able to provide four traditional Daimler limousines. A Rolls Royce hearse and four traditional Daimlers seemed like a good compromise to me.

As the hire drivers were all traveling from the Home Counties and the timings of the funeral would require an early start, my wife Frances suggested that, by way of a thank you for their help, she could produce a round of cooked breakfasts for the five hire drivers when they arrived. The hire companies gratefully accepted the offer, so when the fleet of cars arrived I was left with all the car keys and the job of shunting all the vehicles around our yard into some semblance of order, whilst the drivers made themselves comfortable in my home for their breakfast fry-ups!

With some careful planning beforehand, the fleet of cars was able to turn and line up outside the family's house and even the parking issue at the church was sorted, largely through sheer force of numbers. However, with the family safely settled in the church, which itself was full with mourners, I dared to hope that the worst was over and that I could relax slightly. I was of course to be proved wrong.

I had slipped back into church and from my vantage point at the rear I could make out many familiar faces and well-known local business people. This was all to the good for me – all good local exposure. But then the vicar started giving the address... He started well enough, relating the deceased's early years before moving into his time as a motor dealer in the town and it was at this point that, figuratively speaking, the wheels began to come off.

"Of course many of you will remember Bert's motor dealership at the top of the town. I remember as a young man going to buy a car there, but there was nothing he had that I liked so I went somewhere else." My toes curled. "I'm sure he had some good cars, but I never found one." I listened with dread as the vicar dug himself into a hole deeper than the grave was likely to be.

"Now, I understand that his grandchildren called him Grandpa Sellotape, because he always had sellotape handy for toy repairs or to help with making things for school projects. But I can only hope that Bert didn't use sellotape on any of his cars." By this stage I was silently imploring him to finish and get back to the safety of the prayers, where at least he would have a safe script to work from. But no, worse was to come:

"There is a third hymn, but it hasn't been printed on the service sheet, so we'll just wait while James and his men hand out hymn books instead."

My blood ran cold. Myself and the bearers grabbed piles of hymn books from the shelves at the rear of the church and I hissed to the men to go up the side aisles to hand the books out, as that would at least be a slightly more discreet way of doing it. However, the rule that bearers either forget or just plain don't listen was again proven as one of them walked straight up the main aisle, handing out books and then holding a book up high and waving it around above his head for the benefit of anyone who still didn't have one. He was a retired schoolteacher and I suppose old habits die hard. Meanwhile, I was absolutely fuming – this funeral of all funerals and there we were handing out books because somehow or other we had missed a damned hymn off the service sheet!

Once the vicar had announced the hymn number and the singing was under way I shot back outside and phoned the office on my mobile phone, to check what was on the handwritten draft of the service sheet that the family had given us after meeting with the vicar. There was no third hymn mentioned at all, so wherever the mistake had occurred, it wasn't actually our fault anyway, but that was of no consolation to me at the time.

The burial at the nearby town cemetery went off without further incident and a few days later Paul did tell me that the missing hymn was an omission on the family's part. It was an understandable mistake for a bereaved family to make; it was just a shame that such things can reflect back onto the funeral director sometimes. However, it clearly wasn't as cataclysmic a moment as I'd thought, as that funeral was the first of more than a few which would otherwise have gone to my former employers but instead came to us, with Paul's endorsement.

Having been so preoccupied with the funeral arrangements, it was only much later when it suddenly struck me that the funeral service and the subsequent burial had taken place in exactly the same church and local cemetery as the very first funeral that Paul had taken me out on during my work experience. The circle was complete.

In previous chapters I've mentioned "the other James," my erstwhile colleague at Thomas Broad & Son. It was just two months after that eventful funeral for Paul's family that I invited t'other James to leave Thomas Broad & Son and join me at Fred Stevens Funeral Directors. I was finding things too much to cope with on my own and I decided that two of me – in other words a second, multi-tasking funeral director - would be better than simply having an assistant.

Generally speaking, t'other James and I now deal with the funerals in equal measure and so he too has built up his own memory fund of happenings, incidents and occurrences. Twice now in particular, I have watched as he has very adeptly handled some rather unusual and amusing experiences with families visiting our chapel of rest.

Both incidents occurred whilst we were still in our original, very small premises. A woman had come to see her deceased mother. The daughter was a little bit hippy and way out - in marked contrast to the rest of her family - and it came as no surprise to the other James when, after a few minutes in the chapel, she poked her head round the door and asked if it would be ok to sing some chants in there. James said that yes, of course, that was fine, after which he then had to listen to her warbling away with some form of ritual she had planned.

After a few minutes of chanting, the chapel door opened again, she poked her head out for a second time and with a completely straight face she asked if James could help her with the chanting. Now, by this time, my mother had "retired" from her part-time book-keeper/secretary role and had been replaced by Caroline, our office manager. So, as Caroline was also in the office and unable to get on with any work whilst the deceased's daughter was still in the adjacent chapel, Caroline couldn't resist listening at the chapel door as the reluctant, but ever professional James took the daughter's instructions as to when to chant and what words to use.

Caroline later told me it was all she could do not to burst out laughing as she heard James and his loopy client in the chapel find their rhythm and begin chanting in unison. Our chanting visitor had left by the time I reappeared in the office and Caroline told me what had been going on. James just stood there glaring at the both of us, but Caroline was too busy giggling hysterically to notice. I just wish I had been there to listen to James and our chanting client making sweet music together!

The next viewing incident was as touching as it was hilarious. James was dealing with a family who were, well, colourful, shall we say. The deceased, a gentleman in his fifties, had left behind a son and a daughter. The son was clearly not someone you would want to leave alone in a room with your family silver. But, for all that, he was still a pleasant enough and remarkably polite lad – just very lively and other-worldly, in a way that someone under the influence of illegal substances might be...

His first visit to the chapel of rest, accompanied by his sister and his girlfriend was also a rather lively affair and certainly not as subdued as the majority of viewings usually are. However, it was their second and final visit that was to prove probably our most memorable viewing ever. There was a loud knock on the office door and there once again stood the son, the daughter and the son's girlfriend.

"'Hiya! Alright? How's Dad been? Is he ok?" I pretended to be invisible in our cramped little office and let t'other James handle things with his usual calm diplomacy. James played along well and said "Your Father's been fine. We're looking after him for you" and with that he showed the three of them into the chapel.

"Hiya Dad, we're back. You been ok then? Bellowed the son as he stepped straight up to the coffin and held his father's hand. James and I switched into the "hearing but not listening mode" that we adopted when families were in the chapel, but the tiny size of the office and the fact that the chapel was just behind the door meant there was little we could do other than stand there quietly and listen anyway.

T'other James was the first to hear it – a faint but unmistakeable "ta-ta-ta" noise coming from the chapel. I moved closer to the door and could hear it as well. A constant, rythmic "ta-ta-ta-ta." The mystery was solved when the chapel door swung open and the daughter appeared, asking James if it was ok to put some photographs in her father's coffin.

James stepped forward and said "Yes, of course, that's fine. No problem at all," using the moment to steal a glance through the doorway to where the son was stood by his father's coffin, holding a portable CD player. James' eyes followed the cable leading from the CD player down into the coffin, where the son had placed a set of earphones over his dead father's ears. The noise we'd heard was the sound of the music that was playing on full volume. The son saw James looking.

"We're just playing him some of his favourite tunes. Aren't we Dad? You love this one don't you?"

James shut the door quietly and we had to try so very hard to keep a dignified silence and not just wet ourselves laughing. The lively trio eventually emerged.

"See ya at the crematorium tomorrow Dad." The son shouted back at the coffin before he closed the chapel door. Then turning to James he said,

"Thanks for that mate. You've looked after him really well. He looks really good don't he?"

James smiled gently by way of a reply.

The son, his girlfriend and his sister walked back out into the car park and piled into a battered old Nissan Sunny. As the car pulled away, which in itself seemed a direct challenge to the accepted laws of physics, three different arms were waving out of the open windows and the son tooted the car horn loudly. Beep beep. Then, leaning out of the car window, he shouted across the car park,

"BYE DAD! SEE YOU TOMORROW. BYE!". Beep beep beep.

Not so very long afterwards the son was in the local newspaper. He'd been arrested for being in possession of a controlled substance. Or was it burglary...? I can't remember.

CHAPTER SIXTEEN

Tread Lightly Upon The Earth

"Green burial isn't about doing extra things. It's about what not to do."
Joshua Slocum

"Green" is the new buzz word. Providers in every aspect of our lives are proclaiming their green attributes and the funeral profession is no exception. However, with cremation accounting for 70% of all deaths in the UK, there are challenges in trying to make funerals more eco-friendly.

Meanwhile, we're also seeing a massive growth in more personalised and/or contemporary funerals, which themselves often encompass the green element as an expression of the deceased's values or beliefs. Inevitably therefore, personalised and contemporary funerals have become lumped together under the heading of "green funerals". The ever-growing movement towards funerals which are either green, contemporary, personalized, or more often a combination of all three, requires me as a funeral director to manage huge changes in the way I work and in how I meet the expectations of my clients, many of whom now also have far greater access to information about alternative choices via the internet.

Meanwhile, the challenge for the funeral profession as a whole is that much of the infrastructure we use (crematoria, cemeteries, etc.) as well as the system of administration for deaths in this country (coroners, hospitals, registrars, etc.) is largely based around more traditional assumptions and so somehow funeral directors, as the only private sector part of this jigsaw, must find ways to re-invent ourselves and our relationships with the more immoveable forces around us in order to satisfy the needs of a more demanding public. We are breaking new ground and for funeral directors like me that has inevitably brought many new, interesting and sometimes peculiar experiences with it.

Late one morning I received a telephone call from a gentleman with a wonderfully broad Gloucestershire accent. Not surprisingly, he got straight to the point:

" 'Ello. I need some advice please. I want to know how to measure my wife for 'er coffin. She ain't dead yet but I want to make a start on it. The coffin, I mean."

I decided simply to take him at his word to begin with and explained firstly in very basic terms how we measure bodies, with length (as distinct from height) and width across the shoulders. I asked him if he was intending to construct a traditional tapered sided coffin or an oblong, casket-shaped version. I explained roughly how to decide where on the overall length of the coffin the bend for the shoulders should be located and as the conversation wore on it transpired that his wife was terminally ill and being nursed at home. His intention was to construct the coffin himself and then bury his wife on their own land. I said that, naturally, if he wanted any further assistance at any stage he would be more than welcome to contact us again and with that our conversation ended. I heard nothing more for a month or so.

Then early one autumn evening the phone rang, with one of our local doctors on the line. He was calling from one of his patients' homes to ask if we could come up to the cottage and assist the family with preparing the body of a deceased lady. The family's wish was for her body to remain at home and they wanted our help to bring the body down from the bedroom so that it could then be placed in the coffin that her husband had made for her. Instantly recalling that strange phone call a month earlier, I noted the necessary details. The doctor then mentioned that the cottage was right out in the wilds, which didn't surprise me in the slightest, before giving me directions.

Myself and one of my bearers drove out into the night and armed with a mercifully good set of directions from the doctor we managed to find the isolated cottage without any great difficulty. The husband answered the door and introduced me to a friend of his wife who had been helping to look after her in her final weeks. Everything was starting to click into place now. The friend, herself quite a character and whose father's rather alternative funeral I had conducted a few years previously, had recommended us to this new family. The dead woman's husband was a countryman, woodworker and land owner. His late wife had loved their cottage in the wilds and all that came with it and was determined not only to die there but to remain there afterwards.

My first task was to lay out the body and I will always remember the loving care with which the deceased lady's friend assisted me. Laying out a body – washing it, closing the eyes and mouth to set the features and plugging bodily orifices – is an intimate physical process and I was struck with how this lady's friend didn't once shy away from helping with the procedure, showing not even a hint of embarrassment when we dealt with the very intimate offices that were required.

The husband meanwhile had been having a slight difficulty with the coffin - a difficulty which became all too obvious when I went back downstairs to see the finished article. The coffin was by any standards a work of art, beautifully hand-crafted from cedar wood. The only problem was that it was substantial to the point that he could barely lift it, having been fashioned from nearly inch think wood. As I admired his handiwork he went on to proudly point out the rustic wooden beams in the room that he'd fashioned and installed himself. He was clearly a craftsman with impressive skills.

He went on to outline his plan for the burial itself, in a wooded glade down in a small valley below his cottage. I tried to imagine him and his friends carrying that coffin over such a distance and I pitied them in advance. They already had plans to hold a non-religious funeral ceremony, with assistance from the Humanist Officiant who I had employed to conduct the funeral of the friend's father.

To the husband's relief, myself and my assistant made easy work of bringing his wife's body down the winding cottage stairs with our stretcher and the four of us placed her in the waiting coffin. My work there was finished. I made sure that the husband firstly knew about how to register his wife's death and also that before he and some other friends set about digging the grave, that he would inform the Environment Agency of his intentions, in order to check that there were no watercourses near the grave. Then I left him with his friends at that isolated little cottage.

About a year or so later I met the dead woman's friend again, to arrange the funeral of another of her relatives. I enquired how her friend's funeral had gone, to which she told me that everything had gone perfectly to plan, that they had managed to carry that coffin down into the valley – albeit with plenty of help - and that the ceremony itself was as beautiful as its location. It was satisfying to know that they had achieved all their aims.

My next encounter with a DIY coffin was an altogether simpler affair, but no less poignant in its way. A lady of fairly senior years came to me enquiring about purchasing a cardboard coffin for her own funeral, with the intention of having it delivered to her so that she could decorate it herself. She explained that she was terminally ill and her life expectancy was now measured in months. Her plan was therefore to decorate the coffin and then for us to store it until need arose.

At that time the market for cardboard coffins was considerably more limited than it is now and there were few suppliers, with those few only offering a choice between only plain white or printed wood-effect finish, unlike the vast array of bespoke pictorial designs also available nowadays. As the lady intended to decorate the coffin anyway, the plain white version was perfectly suited to her purpose. She explained she would need to make some arrangements to receive the coffin for decorating and asked me to contact her as soon as it arrived at my premises, by which time she would be able to give me a delivery address. The lady felt that a week would be sufficient time for her to complete the decorating. The coffin arrived two days later, ironically packed in a huge cardboard outer carton that was itself as substantial as the coffin. So much for cardboard coffins cutting down on wastage of natural resources! I phoned the lady to say the coffin was now in my possession and enquired where she wanted me to deliver it to.

She was delighted that it had arrived so quickly and confirmed that she'd made arrangements for somewhere to keep the coffin while she decorated it. The local day hospice where she was receiving care had agreed to have the coffin in their art studio for her to decorate it during her art class! But in a wonderfully ironic footnote, she said the hospice had asked if the coffin could be discreetly delivered late one afternoon after all the patients had gone home for the day. They intended to keep the coffin in the hospice's garden shed when it was not being worked on, to prevent any distress to other patients. That all seemed perfectly reasonable but it didn't explain what they intended to do about the other patients during the art class itself.

I retrieved the coffin from the hospice just under a week later, now resplendent with an orange background representing the evening sky, with a tree in blossom stretching up across the lid and with some of its branches reaching over onto the sides of the coffin. Cardboard coffins are supplied in a universal standard size and I could only imagine the scene in the hospice art and crafts room with this huge coffin either laid out on a table or propped against a wall, with its intended occupant cheerfully painting her design onto it. Her finished masterpiece coffin was in storage with us for only a short time before it was needed again.

It was the wishes of another terminally ill woman, wanting to plan her own green burial, that would lead to my next adventure.

The lady in question, Miss. Parsloe, lived in a sprawling, hillside village on the edge of the district, in a bungalow beautifully tucked away within a large garden. The views across the wooded valley from her lounge window were stunning and it was by this window that the two of us sat in armchairs while she outlined her plans to me. Miss. Parsloe had, via a friend of a friend, been in contact with a young farmer just over the county border in Wiltshire, who was intending to establish a green burial area in one of his fields. To complement the burial ground, he also planned to convert an attractive outbuilding next to his farmyard to accommodate wakes and post-funeral gatherings. He saw himself as being well placed to diversify his farming activities by hosting green burials. Meanwhile, Miss. Parsloe – a spinster with only two nieces as her nearest relatives, intended to make advance arrangements to be buried at the farm and to provide refreshments for the mourners afterwards in the farm outbuilding.

We talked through the arrangements together and agreed a very simple plan – which she intended to then set out to her two nieces. Although Miss. Parsloe had set her heart on a farmland burial, she hadn't actually visited the site and felt it would be a good idea if we went to see the burial site together, so that I could meet the farmer myself and agree the practical details. We fixed a visit for the following week and so, late one summer Tuesday afternoon, I arrived at Miss. Parsloe's bungalow ready to travel with her over to Wiltshire.

Despite my offer to drive she was insistent that as I'd given up my time the least she could do was return the favour by driving us both in her car. I looked at her elderly Vauxhall Nova with a great deal of suspicion but reluctantly decided it would be churlish to refuse her offer to drive. We set off on our twenty five mile odyssey, but it quickly became obvious that the dear lady's grasp of the Highway Code could have done with some improvement and by the time we reached the A419 – the dual carriageway leading across to Swindon and the M4, I was starting to fear for my life. I was a helpless passenger in a car being driven by someone who had no more than twelve months to live and who doubtless didn't have as much to lose as I did!

It was with some relief, albeit short-lived, that we eventually turned (or rather, veered) off the main trunk road back onto B roads and began following the farmer's directions down the winding lanes to his farm. I started breathing again only when we actually arrived at our destination. Still shaking as I got out of the car, I found a very well spoken, overall-clad, young farmer waiting to greet us. With the introductions out of the way he showed us the outbuilding and explained his plans to smarten it up for funerals. I tried to concentrate on what he was saying, but I kept finding my thoughts straying back to a near-death experience with a Ford Transit van, courtesy of my spinster chauffeur's abysmal driving on the journey down.

The farm building itself certainly had potential but I felt that the fact it was accessed through a farmyard may have put off some potential clients. Turning to the subject of the burial itself, the farmer said that although the field was no more than a brisk walk up the slope overlooking the farmyard, the ground in the field was hard enough to drive on, so he suggested that we follow him up the lane in the car. I braced myself for another hair-raising excursion in the ageing Vauxhall Nova, wondering how Miss. Parsloe would cope with having to drive off-road as well this time.

To my utter relief we were required to drive no more than ten yards along the perimeter of the field, before walking the last few metres to a little patch of long grass, dotted with young trees. The farmer told us there was already one other grave there. The farmer explained that according to his researches up to three or four burials would be classed as non-commercial private land burials for relatives or friends, but any more than that and he would be deemed to be running a public burial ground with all the attendant planning issues and requirements for infrastructure, such as access and parking.

The site wasn't much to look at, and whilst of course minimal visual impact is one of the main aims behind natural burial, it was the total lack of any visual impact at all that struck me. It was essentially an untouched area of scrubland at the corner of the field. This would be green burial at its purest – uncompromisingly natural.

What I did notice immediately, as indeed did Miss. Parsloe, was how quiet and undisturbed the location was. The only sound we could hear was the birdsong that carried on the breeze. With the burial site being at the top of a slight rise, I surveyed the views around us and spotted a white horse monument carved into the hillside – a feature that that had always held a fascination for me during my early childhood in central Wiltshire.

Whereas Gloucestershire is famous for the Cotswold Hills and the River Severn, Wiltshire is the county for white horses. There are, or in some cases were, at least twenty-four of these equine hill figures across Britain, with no less than thirteen being in Wiltshire. Most of the white horses were chalk hill carvings, the chalk downs of central Wiltshire making it an ideal place for such figures. Being within sight of that white horse suddenly added a certain resonance to the intended burial site and Miss. Parsloe was delighted when I pointed out the horse to her.

In fact, that made her mind up for her. She confirmed there and then with the farmer that she would be buried there and I began glancing round again making mental notes about the site for future reference. The farmer and I then chatted about contingencies for bad weather, who would actually arrange the digging of the grave and so forth, all the while with Miss. Parsloe listening enthusiastically as two men planned in great detail how they intended to transport her body to a field and bury it!

The time eventually came and mercifully the weather was on our side. It was breezy and grey, but dry and as I drove Miss. Parsloe's woven willow coffin into the field in my estate car – she did not want the formality of a conventional hearse - some of her friends and the husband of one of her nieces carried her willow coffin over to the grave. Inside she was wrapped in the special afghan blanket she had shown me when we had very first met and which since then she had been keeping specially for this purpose.

Her funeral ceremony was a very brief and simple one, held around the graveside, consisting only of readings, poems and words of farewell from the small gathering of people. After that, the little group of mourners walked across to the edge of the field and back down the slope to the farmyard, where catering had been laid on in the newly converted outbuilding. I was left alone by the grave, looking across one last time at the distant white horse, waiting for the gravedigger in his pick-up truck to trundle across from his hidden parking place on the other side of the field.

My next natural burial would be a very different affair, with a cathedral funeral service and my first visit to a beautiful, commercial natural burial site within the land estate of a castle. The family had come to us via recommendation from a Gloucester funeral director who I knew well and who felt that, with the Stroud area's reputation for being artistic and alternative, we would be more experienced and capable of handling alternative funerals than his own company was. In truth I think it was simply that this particular funeral represented something well outside of my counterpart's urban comfort zone and it was one funeral he was anxious to offload.

Knowing what he was like, I was cautious to begin with and asked him whether there was anything else that I should know about, like money problems or difficult family members. As it turned out however, they were a very pleasant family to deal with. The widow was a bit cantankerous in an eccentric kind of way it's true, but not to the extent that either she or her family came anywhere near to becoming awkward clients. The deceased and his wife were both regular worshippers at Gloucester Cathedral and it was there that the funeral was to be held. The number of people expected to attend was no more than an average funeral and so the service was held in the "Quire", the area containing the choir stalls in the cathedral. Back when I was carrying out limousine hire work I had been on hire jobs for other, much larger funerals at the cathedral, but on this occasion the Quire was a surprisingly intimate setting, considering the towering architectural magnificence around us.

Not very long before that funeral, the cathedral's historic cloisters had found new fame when they were transformed into the corridors of "Hogwart's School of Witchcraft and Wizardry", for the "Harry Potter" films. Obviously the film crew had all manner of tricks for adapting the cathedral interiors into a film set and anything that gave the set away as a cathedral had to disappear. Apparently even the halos on the stained glass figures in the cloister windows were painstakingly covered with coloured plastic filter paper to blend in with the surrounding glass and in the preparation of one window, featuring figures of Adam and Eve, the biblical couple in question were given clothes and even had the trademark Harry Potter 'lightning flash' put on their foreheads!

Any funeral taking place there understandably attracts the attention of the never-ending stream of tourist visitors in and around the cathedral. But this particular funeral, with a woven bamboo coffin and an estate car in place of a hearse certainly looked out of place amidst the gravitas and towering grandeur of the cathedral and naturally caused a few raised eyebrows from the English tourists; whilst what some of the foreign visitors must have made of these strange British funeral rituals I can only guess.

After the service, we left the city en route to a natural burial ground in Monmouthshire. When I first met the family they'd expressed a wish for green burial and I discussed the options open to them, which were then (and still currently are) limited by an almost total absence of natural burial grounds within Gloucestershire. I think that was why my funeral director friend squirmed out of dealing with them and passed them on to me. My best suggestion to the family was a site that I was aware of in Monmouthshire, run by a company who worked with landowners to utilise their land for natural burials. It seemed the closest and most attractive option and when the widow pointed out her late husband had historic family connections with the county, she was more than willing to accept my suggestion.

I visited the site whilst still in the throes of arranging the funeral, to familiarise myself with the location, the route and the journey timings. The burial ground, established in meadows which were originally part of the local castle's medieval hunting chase, was split into differently named areas and on that first occasion I took the site director's suggestion as to the best area for the grave.

After journeying down there one Sunday afternoon, I drove up the track from the road and into a parking area enclosed by woodland, where the site director was waiting for me. As he waxed lyrical about the site we climbed some steps set into the bank above the parking area. We reached the crest of the bank, broke through the tree line into the open and there stretched out in front of me was the most magnificent rolling pasture, with far reaching views across the valley below. I could instantly see why he was so enthusiastic about the site. Despite being a damp, grey autumn afternoon the view was truly stunning. I was not only relieved, having recommended a site that up till that point I hadn't actually seen myself, but I was now looking forward to the family's reaction when they would see for themselves just how beautiful the site was.

There were already about a dozen or so burials there but each grave was purposefully unmarked. The site was kept as a working pasture and I could see a flock of sheep further down the slope. The idea was quite simply that relatives could borrow a GPS device to show them where the grave was, but no flowers or mementoes of any kind were permitted as this would spoil the undisturbed pasture. The site management made no apology for this approach and relied on funeral directors to make it abundantly clear to families that this was natural burial as it should be, that their loved one could rest in beautiful pastures but the price of that would be an unmarked grave.

I have no issue with graves being marked with a tree, in time creating new patches of woodland, but I also admired the stance taken by this site's management. This was not a woodland, this was open land with stunning views and for anyone feeling the need to mark the grave or place mementoes on it, then natural burial probably wasn't going to be the best option for them anyway.

After the cathedral funeral we made the hour-long journey down to the burial site. The journey itself was very picturesque, taking us through the densely packed pine trees of the Forest Of Dean before winding down through the river valley to where the ancient town of Monmouth sits. As the family had requested an estate car instead of a hearse, I once more found myself with only my car radio and the coffin in the rear for company. When we arrived at the site the weather was much better than my first visit, with the sun now just trying to show its face. As we carried the coffin up out of the parking area, our little group broke through the tree line and I could hear a satisfying, collective intake of breath from behind me as the family took in their surroundings. My mission was accomplished.

My second visit to that natural burial site, in early 2010, was an altogether more eventful experience – something of a misadventure in fact. Yet again the story starts with a visit from a terminally ill lady who wanted to make advance plans and within twelve months we received the call from her family to tell us that she had died. Arrangements were made for a very simple, civil ceremony at the graveside, but the burial then had to be postponed following the worst snowfall in Britain for twenty years.

I know that up north, where snow is a more regular event each winter, people probably accuse us "southern softies" of grinding to a halt at the sight of just one snowflake, but the west country hadn't seen snow like this since 1982. Eventually the snowy weather departed and the burial was rescheduled, but nevertheless trouble was still waiting in the wings. To start with, the family wanted to speak to the burial site's manager and agree a location for the grave themselves, which was perfectly understandable and their chosen spot was located on the opposite side of the burial site to where my last burial there had taken place.

This little detail didn't cause me any great concern. But because the site was in Monmouthshire I couldn't spare the time for a trip to find the location of the grave as I would normally have done, so instead I was relying purely on my memory of that first burial there a couple of years previously. Unfortunately my memory proved not to be as clear as I thought it was and I was blissfully unaware of just how far across the burial site this latest grave was. Had I known just how far, I would have employed additional bearers for the funeral, but the combined weight of the solid pine coffin, together with its occupant, seemed well within reasonable bounds for the usual four-man team.

My other, much bigger mistake, was assuming that being on a hillside the ground would be well drained. I had debated about getting the bearers to wear wellie boots, which wouldn't actually have looked too out of place on this particular funeral anyway, but making the assumption that the ground would be well drained and firm I decided wellie boots were probably unnecessary. That was a really bad decision, because I didn't realise that under the pasture the soil was still absolutely waterlogged from all the snow....

We arrived at the burial site with the coffin in an estate car, as the deceased had requested. I greeted the family and the civil celebrant, who to my utter my dismay were all sensibly dressed in wellie boots and sensible clothes – unlike me. But, with no choice but to press on, and clutching a pair of trestles for the coffin to rest on during the ceremony, I uttered what was to be a fateful phrase:

"If you could all just bear with me for a moment, I'll just nip over to the grave to put the trestles in place and make sure everything is ready, before we all make our way over."

Nip over to the grave?! More like trudge, squelch, slide and wade for a hundred yards, occasionally stop to rescue my shoe from the mud, try to hop over the small stream that had formed through the middle of a wide dip in the pasture and then hike another fifty yards to the grave. The only good thing about that distance was that the family party couldn't hear my language! I set out the trestles, put out a pair of lowering webs for the coffin and began my soggy exodus back to the parking area, all the while wondering how on earth we were going to get the coffin back across.

The odyssey began. Within seconds the bearers were sinking into the ground with the weight of the coffin and as the distance we walked grew further, so the coffin became heavier for them to carry. The bearers were carrying "under-arm" – shouldering just wasn't an option with the ground conditions. I could hear one of the bearers breathing very heavily and I nudged him to the middle position on one side of the coffin so that I could take his place at the heavier, head end. Eventually, even one of the men in the family had to step in to help as well because the ground conditions were making the carry harder and harder; not to mention too that all of us felt like our arms were about to part company with their sockets from all the weight.

How we ever got to that grave I will never know, but I cannot describe the relief of finally being able to set the coffin on the trestles and have a rest whilst the ceremony started. The ceremony itself was beautiful in its simplicity and mercifully the burial proceeded without further incident.

Whilst the family dispersed, the bearers and I remained by the grave. Some wooden boards had been left for us to put over the open grave to keep the area safe until the gravedigger returned to backfill. Just as we put the last board in place, a lady dressed in an authentically battered waxed jacket and the obligatory suede leather country boots, appeared in the distance, marching towards us. I sent the bearers back on ahead to the car park and waited for the lady to reach the grave.

"I'm surprised you chaps didn't wear wellie boots. It's damned wet down here." She barked cheerfully. "Darn' long way to carry a coffin too. Haven't you got a trolley or something that you could've used?"

I was willing to concede the point about footwear, harbouring thoughts about trench foot as I felt my feet squelching in my shoes.

However I was tempted to give her a blistering reply about the distance the coffin had to be borne to reach this particular part of the site when it dawned on me that she must be the actual land owner, rather than someone from the burial site management company. So, deciding discretion was the better part of valour, I decided on a more gracious reply. As the two of us walked back to the parking area she did talk of her ideas for having a pony and cart to transport coffins across to the more distant parts of the burial site and I agreed with her that would be a really nice idea. I was tempted to suggest a decent cart horse instead of just a pony, then we could *all* have been saved a long walk!

From a funeral director's point of view, the key to arranging and conducting contemporary and alternative funerals is being able to gauge both the spirit and the aesthetic of each individual ceremony and then plan accordingly. For some funeral directors that can mean a big step out of their comfort zone, leaving behind the security of the conventional black hearse in favour of a normal estate car, or leaving the frock coat and top hat behind in favour of a normal business suit, for example (or sometimes just a sensible waterproof coat and a pair of wellie boots...). Personally, this has never been an issue for me; in fact I would have felt rather more self-conscious if I *hadn't* blended in with the aesthetic of some of the funeral ceremonies I've been involved with over the years.

However, the challenge is far more complex than just stepping away from the norm. The growth in contemporary funerals has brought with it a bewildering array of options: everything from natural coffins through to alternative modes of transport such as horse-drawn carriages, vintage lorries and motorcycle sidecar hearses. The challenge is to avoid, as so brilliantly described by sociologist Dr. Tony Walter, of Bath University, falling into the trap of "mass produced individuality."

What I believe Dr. Walter was getting at is that things such as eco-friendly coffins, alternative forms of transport and any one of a number of other options available nowadays, such as fireworks or dove releases for example, should only be seen as accessories and not an end in themselves. There seem to be some funeral directors who are happy to market such items either because they think that's all they have to do to be able to accommodate contemporary funerals, or because they see the sale of such items as a means of increasing earnings whilst still being seen to be forward-looking and modern.

The fact is that an alternative coffin, with alternative transport and accompanied by say, a dove release, will not create a personalised funeral. Far from it. The challenge is not just to provide tangible items like contemporary coffins or alternative transport, but to work with families to add meaningful personal touches that often cost nothing more than a little time and effort and then weave those elements into a well thought out funeral ceremony.

There are also a great many practical considerations to be borne in mind when using alternative transport – particularly horse-drawn hearses. For one thing I make it a personal rule never to ride on the carriage. There is an art to clambering up onto, or down off, one of those carriages in a dignified manner and having to do that a couple of times en route, for either leading away from the house or on arrival at the church, is considerably easier if you are already near the ground in a car. In our area we also have to be mindful of whether the horses would be able to cope with the hills, whilst cattle grids are also a constant consideration for us in some parts of our district!

As a more low-key alternative my company offers a horseback outrider, usually for instances where the deceased had an interest or connection with horses, but when a horse-drawn hearse would be either too ostentatious or just not practical. "Harry," the horse we currently use, was carefully picked for his excellent temperament and his almost bomb-proof attitude to traffic and all the other things that can spook a horse. Our rider herself is a former professional show jumper with the experience and ability to deal with the horse if he were to take fright at something. It all has to be thought of.

I used to be absolutely fascinated by farms and tractors when I was a kid in rural Wiltshire and my mother often says my first words were from the name of a well known tractor manufacturer: Massey Ferguson. She probably would've settled for a simple "Mum," but such is life!

So it's appropriate then, that the very best example of alternative transport I've seen involved an old Massey Ferguson tractor. It was the funeral of a local farmer's wife and her flower garlanded, woven willow coffin was taken from their farm, down through the village to the church, on a tractor and trailer. Just after the Second World War, as a young, newly married couple, this woman and her husband had bought a farm in the hilltop village where they were to spend their entire married life. I was shown a photo of the two of them taken at that time – a touching snapshot of a young couple very much in love, smiling for the camera, at the start of their new life together. The day they moved into their farm they arrived on a tractor with all their belongings piled up on a trailer. And that was how the farmer wanted his wife to leave their farm for her funeral.

The couple's children decorated the trailer with flowers, they placed their mother's willow coffin upon it and with their father at the wheel, we all walked together in procession, down through the village, accompanied only by the gentle putt-putt sound of the little veteran tractor's engine.

Just as he'd done as a young man all those many years ago, when he'd driven his young bride to their new home, now the elderly farmer drove his wife away from the farm on a tractor again, for the very last time. It's being part of moments like that when you're reminded of just what a privilege it is to be a funeral director.

CHAPTER SEVENTEEN

Private Sector – Public Service

"Is the system going to flatten you out and deny you your humanity, or are you going to be able to make use of the system to the attainment of human purposes?"

Joseph Campbell

It was late on an autumn afternoon when the phone call came. I answered and found myself talking to someone at police control. Could we attend (never "go to" with the police, always "attend at") an address in a village a few miles south of Nailsworth, to remove the body of a man whose name the control operator then gave me. Glancing at the office clock to log the time of the call, I realised that by the time we had got to the address and performed the removal, the body would have to be taken to the public mortuary as an out of hours admission. So after a quick discussion with t'other James we decided that he would have time to do the removal with me before having to get back home in readiness to go out with his wife that evening. I would then transfer the body on over to the mortuary myself later in the evening.

We both knew the village where we were going, but the operator at police control had given me some directions to the precise address and I knew from those directions that the house was going to be fairly isolated. My instincts told me something might be afoot. By the time we got down the lane where the house was, it was already dark. We came round a bend in the lane and through a hedge on the right the bright flash of blue and yellow police car markings reflected in the headlights of our private ambulance. Stopping at the foot of the long driveway, we could see another two police cars parked further up the drive, along with a small white van.

"I thought there was something up when the call came in." I remarked to the other James. "That's that scene of crime officer woman's van. I saw the same van at that house in Stonehouse a while back and she was there on that one."

"Stonehouse?" asked James.

"Yeah, the woman whose clothing touched the stove and caught fire when she was cooking her tea."

"Oh, yeah, that one."

"You'd better go and see what's what and I'll turn round and reverse up."

T'other James got out and disappeared into the gloom. No sooner had I turned the ambulance and parked up than my phone rang. The office phones were diverted onto my mobile and it was one of the district nurses based at the surgery next door to our office. Another removal: one of their terminally ill patients had died at her home in Nailsworth. I explained that we were out on another removal and asked the nurse whether she thought the family could cope with a ninety minute delay before we could get there. I could hear her having a discussion with the family in the background before she came back on the phone and told me that the family weren't at all anxious and would be quite happy to wait. I jotted down the details and finished the call. As if on cue, James reappeared out of the darkness, opened the passenger door and slung his clipboard back onto the seat.

"You were right about it being a dodgy one. It's a suicide. He's done it with a chainsaw."

"Nice." I replied. "You don't do *that* unless you mean it."

We threaded our way through the police cars with the stretcher trolley and made our way round the side of the house to where the body lay under a blanket. Two of the police officers aimed their torches onto the body in an attempt to give James and I enough light to see what we were doing, but torchlight was barely enough to work in.

"Be bloody careful where you put your feet" James muttered to me, "there's blood all over the...whoa!... path." With my eyesight gradually adjusting to the dark I just about managed to see James nearly slip over. Regaining his balance, he finished his sentence, "...There's a gaping injury to the neck."

"That's all it would have taken to do it." I replied thoughtfully, momentarily imagining the scene. "He would never have managed to decapitate himself, much less needed to anyway."

It's a well recorded fact that the majority of people intending to take their own lives will use the method most readily available. Yes, there are times when people show almost ghoulish ingenuity in establishing a method by which to die, but usually the would-be victim is concerned only with ending their torment and the method itself often becomes nothing more than a practical consideration based on convenience and availability.

Moving in and out of the small pools of light from the police officers' torches, we opened out a body bag on the stretcher before each getting a firm grip on the body so as to be able to lift it up off the path and into the waiting bag in one single manoeuvre, to prevent everything getting soiled. The moment the body was zipped up and strapped onto the stretcher trolley the two police officers unhelpfully switched off their torches and returned to whatever they had to do, leaving James and I once more plunged into stygian gloom as we navigated our way back round the house to our waiting ambulance.

With the body temporarily deposited in the mortuary at our funeral home, James headed off home whilst I took one of our part-timers out on the second removal. Arriving at a little cottage in a hidden corner of the town, I found the deceased lady's husband and son waiting patiently for my arrival. The moment I stepped into the lounge of the property I saw the lady's freshly washed and dressed body lying in a hospital bed that had been installed in the lounge, to enable her to be amongst her family during the final weeks of her life. Around the bed was all the usual medical and nursing paraphernalia that always accompanies terminal illness.

Her husband was more concerned about how we would remove her from the cottage:

"I'm afraid that in the last stages of her illness it was difficult to keep things clean, but the district nurses have washed her and put clean clothes on, so I hope that'll be okay. Now, we'll move some of the furniture and I'll put the outside light on as well so you can see your way down the garden path. Will you be able to manage alright? I'm sorry this is so difficult."

"This really won't be difficult for us" I gently assured him. "This is quite straight forward for us. We've encountered far worse situations..."

Funeral directing is one of a very specific group of professions that between them are responsible for handling and administering the approximately 600,000 deaths that occur in England and Wales each year. The funeral profession is also the only private sector profession directly involved with death in this country and because of that, many strange contradictions surround our role. Let me briefly explain the professional structure of death in this country, before explaining some of the contradictions.

We start at the top with H.M. Coroners. Her Majesty's Coroners are independent judicial officers in England and Wales who are bound by laws which specifically apply to coroners and the holding of inquests. Although they are appointed by, and paid for, by county councils or the relevant local authority for metropolitan districts, they are not local government officers, but hold office under the Crown. Coroners must be either a solicitor, barrister, or doctor (in some cases they are qualified in both fields) of at least five years standing. The requirement to be legally and/or medically qualified reflects the role of a coroner: to investigate deaths that are sudden, unexpected, occurred abroad, were suspicious in any way or happened while the person was in police or prison custody.

We then have the Registrars For Births, Deaths & Marriages, who are responsible for registering every death in England and Wales. For a death to be registered, the registrar requires either a Medical Cause of Death Certificate signed by the doctor attending the deceased during their last illness, or a coroner's certificate stating the cause of death.

Burial or cremation can only take place after the death has been registered, unless the death is subject to a coroner's inquest, in which case the coroner will give permission for burial or cremation to take place.

The next step down comprises of the medical profession. Either the deceased's G.P., or the relevant hospital doctor, will issue the Medical Cause of Death Certificate which will enable the death to be registered. Conversely, if the death is sudden and unexpected, then it is also the doctor's duty to notify the coroner of the death.

Having dealt with coroners, registrars and doctors – the main characters in death's drama, we then add in the supporting cast: mortuary technicians, hospital staff and crematorium/cemetery staff. These people are responsible for regulating either the release of the body or its final disposal.

And finally the principal cast member, the main protagonist of the story if you will: the funeral director. By now, I hope that the previous chapters have given you some insight into what we as funeral directors actually do and what our role is.

Amongst all the various professionals that I have listed above, funeral directors have a number of unique distinctions: we are the only ones involved in every aspect of death – administrative and practical; we deal daily with doctors, registrars, coroners, the police, etc.

As funeral directors, we not only have our own tasks to perform, but – if we are performing our role properly - we also have to act as guides, translators, agents and contractors for our bereaved clients. As such we have to know about the criteria governing the issuing of medical death certificates, coroner's procedure and the law surrounding it, the legalities and procedures surrounding the registration of deaths and the law of burial and cremation. All of this gives us a perfect overview of the entire structure. And there lies the first of the contradictions I mentioned earlier.

Despite our all-encompassing involvement with the various stages of professional intervention in death, because we are a private sector profession we are the only death professionals who are non-statutory. You cannot legally avoid doctors, registrars and coroners. In turn, when taking custody of the body and carrying out the chosen method of its disposal you cannot avoid mortuary or nursing staff, or crematorium/ burial authorities. But you do have a choice as to whether you use a funeral director.

We are not a legal requirement. If you have the time, the inclination, the practical resources and the mental fortitude then there is nothing to stop you organising the funeral yourself and disposing of the body. But only roughly two percent of the population choose that option.

The fact of the matter is that funeral directors are the only death professionals who the bereaved actually *choose* to appoint. Admittedly, it's not much of a choice, because over ninety percent of the population, when faced with a bereavement, will automatically appoint a funeral director rather than contemplate carrying out the funeral themselves. That could be seen as rather un-empowering for the bereaved. However, the bereaved still exercise their freedom to choose *which* funeral director to use and so once again the balance of power is restored back in favour of the bereaved.

I once attended a lecture given by a university graduate funeral director, entitled "Consumer Behaviour In Funeral Service". His aim was to identify how bereaved families set about choosing a funeral director. Such information is, to the funeral profession, on a par with learning how to turn base metals into gold was to medieval alchemists. Unfortunately, the results of his research proved equally as mysterious as alchemy. He didn't come up with any great revelations and couldn't identify any reasons beyond those we all knew already: that it's a combination of personal recommendation, previous family precedents, the funeral director either being personally known to the family or known through being well-established locally, location in relation to the client, the deceased having purchased a funeral prepayment plan with a specific funeral firm, or just a random choice at the time of need.

Even when the bereaved choose which individual firm of funeral directors they want to appoint, their chosen funeral director will still be at the mercy of organisations or officials who, because of their professional designation, hold all the cards. The law is the law and of course as funeral directors we understand and accept that. Indeed, at times funeral directors actually have to *defend* the role of doctors, registrars and coroners to bereaved clients who may be anxious, impatient, angry or mistrustful towards the various authorities involved.

Funeral directors know better than anyone that there is a very good reason why deaths must be properly certified and registered: to prevent foul play and subsequent disposal of the evidence;

We are sensitive to the fact that many people find the thought of a loved one's body being subjected to a post-mortem deeply distressing, but we also know that the results of coroners' investigations, together with the findings at inquests, can often result in recommendations being made that will help to avoid similar tragedies occurring in the future;

We know that cremation has laws and regulations to both prevent wrongful disposal of bodies and to address environmental issues arising from cremation.

But take one step below the statutory, legal level of coroners, registrars and doctors and frustration lurks, in the largely hidden world of hospitals, mortuaries, crematoria and cemetery administration. Our working days can often consist not just of appointments with families or conducting funerals, but with chasing round hospitals, coroner's offices, mortuaries and crematoria to collect and deliver bodies and paperwork, all within a timeframe of often hopelessly inefficient and inconvenient opening times.

Despite the fact that funeral directors all operate a twenty four hour service, on a daily basis we have to deal with offices and authorities that only operate between 9.00am and 5.00pm at best, and we are constantly having to squeeze non-existent minutes out of the day to ensure that essential cremation paperwork is delivered to the crematorium in time for medical referee deadlines, or removing bodies from hospitals and mortuaries within restrictive opening times, all at the urging of anxious relatives who are either eager to view the body or just want the peace of mind of knowing that their relative is in our care, rather than in an impersonal hospital mortuary.

Alternatively we may find ourselves trying to explain to clients why the homemade music disc they have specially recorded and which works perfectly well on their own CD player at home will not be accepted by the crematorium staff, both for copyright reasons and to avoid the danger of the disc not playing properly on the crematorium's CD player.

This is where crematorium staff for example – as members of the supporting cast in this whole drama – can come into their own. Ultimately they are each bound by the requirements and regulations of their particular job or organization and all too often – through no fault of their own - they are required to work and act without knowledge or understanding of how their individual role fits into the larger jigsaw. It is the pieces of this jigsaw which funeral directors are routinely required to assemble on behalf of their bereaved clients.

Most of the supporting cast we encounter are accommodating, co-operative and genuinely willing to play their part as well as they can. But inevitably there are a proportion who are not prepared to extend themselves any further than the narrow requirements of their particular job description. Indeed, in my twenty five years I have encountered more than a few who would not even extend themselves that far.

The whole system of death administration only ever gets tighter, but as is often the case with "progress", it doesn't necessarily get any more effective. There are strict laws and regulations governing the certification of death, cremation and burial in this country and yet the funeral profession, which handles 98% of these deaths, together with that 2% of the population who choose the d.i.y. route – are completely unregulated.

Funeral directors are the only people with an intimate and comprehensive understanding of the system of death *in its entirety* and the state does well to at least ask our opinion on how the system should be structured and run. The state does not always give us much of a say but even so it still relies on us to perform its funerary functions because it acknowledges that only we, the funeral profession, actually know what to do.

The most obvious examples would be the M.O.D. military repatriation contract, currently held by the specialist division of a family-owned London funeral company, or the family-owned funeral firm in North London appointed to assist with the funerals of the Royal Household.

So, ironically, the state can send armies abroad to fight wars, or it can bring the country to a standstill to perform, with unrivalled pomp and ceremony, the burial of royalty, but in amongst all of that it also still relies on the technical expertise of funeral directors.

Of course these are very unique and individual examples of the profession assisting the state. But at a local level, across the country firms of funeral directors just like mine are contracted by local authorities to carry out functions such as, for example, Environmental Health Department contract funerals for those who die with no known relatives. This is the modern equivalent of the pauper's funeral, a concept which has never really gone away.

Another role that considerably more funeral directors are involved in is the removal and transfer of bodies on behalf of H.M. Coroners. Arrangements vary from county to county, with specific contractors being appointed in some counties, whilst in others more informal arrangements exist, where either the nearest funeral director to the scene, or the one chosen by relatives if they are present, will be asked to carry out the removal on behalf of the coroner.

Regardless of whether there is a designated contractor or whether individual firms are called ad hoc according to proximity, these removals are always carried out below economic break-even point. We have to accept the risk, in turning out to what could be an unpleasant scene and doing all the hard work, often during antisocial hours, that another funeral director may still end up performing the actual funeral. On the other hand, whether it be bidding to take on a formal contract or just turning out ad hoc when asked, if we don't respond when asked then a competitor most certainly will, with the risk that they might gain the funeral if the relatives become aware that they were the company who responded. Either way, local authorities who ultimately pay for body transport services as part of the coroner's budget, will have private sector funeral directors recovering and transporting bodies for them for very little cost.

But leaving aside the politics of transporting bodies for the coroner, this aspect of my working life has certainly added some colourful experiences to my career. After twenty five years, there is little that I come across in the course of my work that surprises me anymore, but one of many things that I originally found surprising was just how many funeral directors are actually quite squeamish. They will invariably do what they have to, and do it properly, but not necessarily by choice!

I was just the opposite, having spent of the first few years of my career in eager anticipation of different types of coroner's removal. This might seem a very bizarre admission, but these situations were all part of the job as far as I was concerned and I was anxious to experience every last thing that funeral work could offer. In those early years there was undoubtedly an element of morbid fascination too, but even then it was also a desire to learn *how* these things are dealt with: how *do* you deal with a dismembered body, or a decomposed body, or a burnt body?

Back when I was a young employee at Thomas Broad & Son, I remember having a phone call from one of the company's secretaries who was covering the phones that night. Normally at gone midnight I would have been far from chuffed at being asked to turn out on a removal, as no matter how many hours sleep I lost I would still have to put in a full day's work the following day. But the blow was softened by what she had to say:

"The police will be there waiting for you. They said could you bring a knife, or something, with you. He's hanged himself and they haven't cut him down yet."

The address where the death had occurred was not far from where I lived, so Stan, the bearer on call with me that night, picked me up en route to the address. Whereas we would normally start at one end of the road and count the house numbers, on coroner's removals like this we simply looked for the house with a police car outside.

I stepped through the open front door of the property and was met by one of the policemen. I checked the details I had been given about the deceased: full name, age, etc. and then asked to be shown where the body was so that I could plan the removal.

The police officer told me the body was upstairs and left me to go and look for myself. I headed up the stairs and as I got near the top I instinctively turned my head to look over the banister rail that separated the landing from the stairwell. This wasn't my first hanging and I knew what to expect, but what I wasn't expecting as I looked over the banister was to suddenly find myself staring straight into a pair of bulging eyes peeping back at me over the rail.

The shock made me wobble on the stairs for a moment, before I regained my balance again. The man's body was suspended by a rope tied to a beam through the open loft hatch and he had then simply folded his legs beneath him so that all his weight was on the rope and his neck. There was nothing unusual about any of that, but what threw me was simply the shock of him appearing to be peeping at me over the banister rail because his body was still suspended.

There is a world of difference between judicial hanging, as used for capital punishment (where the aim is fracture dislocation of the second and third vertebrae in the neck), as opposed to suicidal hanging, where death is usually caused by asphyxiation or vagal inhibition (stopping the heart by stimulation of the vagus nerve in the neck, caused by the pressure of the ligature). This manner of dying is not necessarily painful, but can often appear ugly for anyone unfortunate enough to discover the body, even for unwary undertakers.

Whilst there is nothing remotely unusual about people taking their own lives in their homes, it says much about modern society that even in a small town like Nailsworth someone can die in their home and lay undiscovered for some time. Two of the most decomposed bodies I have handled both resulted from deaths that occurred not in secluded woodland, or in isolated properties, but in houses in the middle of busy residential estates. Some years ago we were called to a semi-detached house in the middle of a large, 1960's housing estate in Stroud. It was a Saturday evening and the two unfortunate police officers who were called to the scene were determined not to have anything more to do with the body once they'd established there were no suspicious circumstances.

Outwardly there was nothing to set the house apart from any of the neighbouring properties, but inside it was utterly squalid and neglected. The interior of the property reminded me of a horror film set, with every room filthy and piled with rubbish, sack cloth stapled around the window frames in place of curtains, a carpet so decayed that it lifted with the soles of our shoes and everywhere the telltale buzzing of flies.

The sole occupant of the property, an elderly man, had died whilst sat on the sofa in his lounge. The police had reason to believe that he might have been dead for up to three months and judging by the very advanced state of decomposition of his body, still sat slumped on the sofa, I was inclined to agree. We brought the stretcher in, placed it in front of the dead man's feet and laid out an open body bag on top. Taking a side each, my bearer assistant and I were then able to dislodge the fragile body from its resting place on the sofa and gingerly lift it forwards and down onto the waiting stretcher. Although I prevented my clothing from becoming soiled, the shoes I was wearing had thick grips moulded into the soles and bits of the decayed carpet became so wedged into them that I never could get the smell off those shoes again and had to dispose of them.

I happened to be at the public mortuary some time later when the undertakers who held the council contract were there to remove the man's body again. I suspect there were few people, if any, to join the Environmental Health Officer and the undertakers at the man's funeral.

The second incident occurred some years later, on a council estate in Nailsworth. T'other James and I were somewhat surprised to see a police riot van parked outside the house, in addition to the usual police cars and once again the police were determined to stay outside the property, well away from what lay within. Some of the officers present told us that a group of their colleagues were on a training exercise and when the body was discovered they were diverted to the scene on account of being close by, which explained the presence of the riot van.

This time around the inside of the house was no more grubby and unkempt than we would expect for an elderly gentleman living on his own and once again his body was laying on the sofa where he had died. It was suspected that he'd been there for at least two weeks, in the summer heat. The police officers were quite content simply to stay out of the way in the overgrown garden and watch James and I through the lounge window. As we lifted the body into a body bag we checked his pockets for any personal belongings, but all we found were a hundreds of maggots. We could hear a collective cry of disgust from our uniformed audience outside the window.

As we carried the stretcher out of the house I saw one of the police officers grimacing at the sight of the covered-up body.

"I really don't know how you guys can deal with these jobs. And bloody maggots too. Yuck! Just my luck to be on one of the closest units to the scene when this call came in." The officer complained.

"Downer." I replied sympathetically. "To be honest, I'd sooner be dealing with something like a road accident, rather than a decomposed case. At times like this I bet you wish you were in Traffic Division too, don't you?"

"I *am* in Traffic Division!"

"Oh."

When we arrived at the public mortuary at Gloucestershire Royal Hospital, Nicky, the technician, immediately noticed the pungent smell coming from the stretcher as we wheeled it in.

"What've you brought us then?" she asked cautiously.

"Mr. Arthur Hughes….and friends. Thousands of them."

Nicky grimaced and her shoulders slumped with dismay. She knew exactly what I meant.

CHAPTER EIGHTEEN

Silent Running

"Nothing gives one person so much advantage over another as to remain always cool and unruffled under all circumstances."

Thomas Jefferson

Funeral directing, one could argue, is not dissimilar to any other form of event management, in so far as it will always entail a disproportionate amount of work in relation to the actual event – the funeral. Even something as minor as just washing the hearse and limousine can take an hour or so, which is often about the same duration as the average funeral service. But leaving aside all the myriad practical tasks that the funeral director and his staff have to perform, there are a great many other more unexpected, human challenges that we face in trying to ensure that each funeral runs smoothly and according to plan. Think of it like a duck on a pond; seeming to glide effortlessly across the water's surface, all the while his little webbed feet paddling furiously under the water to keep him moving.

For over twenty years I have had to summon up the energy to show polite amusement in response to hackneyed remarks like "Dead end job", "He's the last man to let you down" or "You'll never be out of a job." However, the equally tiresome "It's not the dead you've got to worry about – it's the living!" does nonetheless neatly sum up the challenge of being a funeral director. The entire process of arranging and conducting a funeral is fraught with human challenges, right from that first moment of contact with the newly bereaved.

The immediate stages after a death has occurred can be so traumatic for the newly bereaved that we, as funeral directors, can never assume that the family will be able to think in a normal, rational manner. However, particularly when the death has occurred at home, the vast majority of families are simply grateful for the presence and attention of an experienced professional to guide them and - where appropriate - take charge of the situation. Well, normally they are...

On one occasion, on a rather tatty local estate, I had arrived to remove a deceased gentleman from his home. As we pulled up outside the house I was met with a small group of family members leant against the garden fence, all smoking cigarettes and all with faces fixed into the same hard, unwelcoming expressions. My heart sank. Momentarily I reflected back on those innocent times in the early days of my career, when I'd been to a similar kind of estate and that incident with the garden spade had left me in fear of being eaten alive. Now, with many more years experience behind me, my only real concern was whether I would have any trouble getting paid for carrying out the funeral.

I got out of our ambulance, said hello to the little group and let one of them lead me to the front door. As I walked up the path I began to feel a glimmer of hope because, in fairness, the house did actually look considerably better kept than most of the neighbouring properties. Admittedly the welcoming committee wouldn't have won any awards for friendliness, but then I was there as the undertaker and who has an undertaker round to their house as a matter of choice? No, the house was tidy, the obligatory half-dismantled car in the front garden was conspicuous by its absence and not only was there a lawn, but it even looked like it had been cut in the last twelve months. Things were looking good.

But barely had I set foot inside the tiny hallway than any charitable feelings I was beginning to develop quickly dissolved. The dead man's wife appeared as if from nowhere, positioned herself nose to nose and toe to toe with me and launched into a shrill tirade about what she would do to me and my colleague if we so much as even laid a hand on her husband's body. I took a deep breath to counter my irritation and listened patiently, ensuring that my body language and facial expression looked attentive and non-adversarial.

When she finally ran out of steam and let me speak, I calmly assured her that we would do everything possible not to hurt her husband. I picked my words carefully, so as not to promise anything I couldn't deliver, because I instinctively knew that if I was getting this kind of verbal bombardment from the deceased's wife then Sod's Law was bound to dictate that her late husband would prove to be a heavy body that wouldn't be easy to lift and handle anyway, let alone whilst having to give the appearance of treating his mortal remains with the utmost delicacy. Sadly my instincts would soon prove correct - her husband was indeed a large and solidly built man.

I kept one eye on the other members of family to sense whether they were going to be difficult as well, but it soon became apparent that they were going to keep their silence and let mother do the talking – partly because they were no doubt lost in their own grief and partly because their mother was clearly intent on doing all the talking anyway.

Having let the wife say her piece I asked if there was somewhere we could sit down and have a brief chat. With the two of us now sat in the lounge, a small audience gathered around us and with the all-pervading cigarette smoke already making my eyes water, I slowly went through my usual initial questions, checking the deceased's full name, his age, the name of his GP, asking the wife whether the doctor had said anything about issuing a certificate, before explaining about needing to meet again to discuss the arrangements. This little interlude finally achieved my aim of calming her down somewhat and went some way to convincing the newly-widowed woman that I was in fact on her side.

We performed the removal in exactly the same way as usual – with as much care and gentleness as physical constraints would allow. We are always sensitive to the fact that many families still invest a degree of feeling and personality towards the body of their loved one and on that removal we were, as usual, fully aware that the *body* we were handling was still *Arthur* to his wife and *Dad* to his children and we treated him as such. But none of that exempted us from the fact that his body was heavy, that the stairs were tight and that were awkward corners to negotiate on the way to the front door; but what appearance of gentleness and care we could muster seemed enough to convince the wife that her Arthur was in good hands after all.

When we met again the following day to discuss the arrangements – an appointment I had been dreading - the wife was considerably more friendly and firstly apologised for the verbal bombardment she'd given me when we first met, before thanking me for my understanding and patience. To my eternal relief she also turned out to be what I call "old school" – so ingrained with a dislike of being in debt in fact, that she insisted on paying for the funeral before I'd even finished making the arrangements.

The woman came from solid Catholic stock and with her relatives dotted around various churchyards in the area I would, in later times, often encounter her tending various graves when I was out doing my rounds, if not also bump into her in more mundane locations like the queue in the local Post Office. Ironically, considering how our first meeting went, we have actually become good friends now!

Having got past the moment of initial contact with a newly bereaved family, the next and arguably largest hurdle is the meeting to discuss the arrangements. Many are the times when I have been pleased to find that the deceased has left very specific instructions, only to then listen in dismay and frustration as one member of the family, invariably acting on a personal agenda, will talk the rest of the family out of following the deceased's instructions. This can be particularly frustrating when the deceased has specified arrangements that may be considerably easier for us to accommodate than anything the errant relative wants to replace them with.

Worse still are the times when the deceased's wish for a more personal or contemporary style of ceremony is disregarded on the grounds that Auntie So and So might be offended if the funeral isn't kept traditional and conventional. Ultimately of course, such matters are for the family to decide and as the funeral director I must simply bite my tongue and follow the instructions given to me by the person arranging the funeral.

However, such situations are no more than minor irritations when compared to the dreaded "split family" scenario – every funeral director's nightmare. With the marital break-up rate in the UK as high as it is, this in turn has ramifications for the nation's funeral directors, who then have to deal with estranged relatives who might at best be daggers drawn, or at worst wanting to use the funeral as a blunt instrument with which to bash the opposing side of the family over the head with.

I am legally required to carry out the instructions of the nearest surviving relative or executor and that has to be my default position. However, there are always two sides to a coin and even though I have to act for my primary client I still always try to remain neutral and not let myself be drawn into taking sides.

Taking a firmly neutral stance has meant I have been forced into some dreadful situations, for example having to tell a young man that I was not allowed to let him to see his own mother in the chapel of rest on the instructions of his step-father, or telling an elderly couple that their daughter-in-law was threatening not to allow their son's coffin to be brought into church for his funeral service – a situation that was thankfully resolved, albeit only twenty fours before the funeral took place. Another, more bizarre situation saw me having to re-arrange not only the date, but also the location of a funeral so that a group of very unwelcome relatives would not have been able to gatecrash the service.

So how do I cope with such situations? Well, by never taking sides and trying as far as possible to treat all parties with equal respect and consideration, I am often able to use my neutrality to negotiate a degree of compromise between the warring factions. It's not for me to resolve my clients' family politics, but I will at least always try my utmost to encourage some degree of co-operation to ensure the funeral itself proceeds with dignity and allows all the various parties the space in which to grieve.

However, if all else fails, or if the family situation is just too intractable for any compromise to be reached, then I can only rely on my sense of professionalism. Ultimately I can do only that which the law allows me to do: to carry out the wishes of the person arranging the funeral and having done that I have to walk away and leave the family to their own devices. That may sound very hard, but these situations are never as clear cut as they appear and if I allowed myself to be drawn into refereeing family conflicts I could unwittingly end up making the situation even worse, even by starting out with good intentions.

The average length of an arrangement meeting is about an hour and a half, although I have sat with some clients who have made instant decisions and the meeting has taken just half an hour or less, either because the client knows exactly what they want or simply because they do not want to engage with the process for a moment longer than is absolutely necessary. At the other extreme, I have sat with families or individual clients for three, maybe even four hours, on occasions. Those are the kind of meetings where I am likely to end up with very few decisions being made and barely half the arrangements agreed, even in principal. Such funerals tend to *evolve* rather than be arranged.

With one such meeting, although the family themselves were lovely people, they were just, well, quaint. It took nearly half an hour of discussion with the deceased's two sons and a daughter-in-law just to establish their preferred date for the cremation:

The daughter-in-law was first to put her two penn'eth in:

"I've looked in my diary and of course we're away next week and then the week after that Hugo is sitting his exams. We simply can't expect Hugo to concentrate on his exams if we've also scheduled his grandmother's incineration for the same week!" She declared, clearly horrified at the thought of her youngest son having to endure such a distraction during his exams.

Her husband, the eldest of the deceased's two sons, had other members of family in mind:

"Well the following week won't be any better because Annabel will be in South Africa."

That prompted his brother, the younger son, to speak:

"Oh golly! I hadn't thought of that. I suppose Simon will be wanting to come back over with her too, wont he? He did so get along with Mummy. He would hate to miss the funeral. We simply can't leave Simon out," he remarked.

The daughter-in-law, noting the menfolk's respective observations, continued the list of family members to be allowed for:

"Now what about Dougal and Penelope? They'll have to travel down from Scotland, so it would be better if it was the Monday, so they can use the weekend to make the journey." She suggested.

Her husband threw a spanner in the works:

"I can't do the Monday darling, I'm due in London for a meeting. Could Dougal and Penny manage the Tuesday?"

It became immediately apparent that as well as throwing a spanner in the works, the eldest son had dropped a clanger too:

"She hates being called Penny." His wife replied tartly, glaring back at him. "You know that! You remember how worked up she got when Hector kept calling her Penny all the way through lunch on Mummy's birthday last year."

"Ah, yes, I'd forgotten that lunch." Her husband replied ruefully, as if recalling some dreadful event, like telling a child the family cat has been run over. "Wasn't that when Annabel and Dougal put their foot in it by mentioning about Mummy having to go into a home?"

The second son, seizing a chance to talk about anything other than his mother's funeral arrangements, piped up again:

"Was that at Mummy's birthday lunch? I thought *that* all blew up at Easter."

I decided it was time to drag the conversation, kicking and screaming, back onto the subject of the funeral.

"So, we're looking at maybe four weeks Tuesday then?"

The eldest son decided to speak up again.

"I think it'll have to be. Poor Mummy, having to wait above ground for so long." He stared reflectively at the floor, then his head flicked up again. "Oh, but hang on darling," he said ominously, glancing at his wife. "Isn't that when you and I are booked to be away on Hugh and Cressida's canal barge...?"

I bit my tongue, dark thoughts swirling in my mind. Most depressing of all was that as I glanced down at my arrangement sheet I knew I was still less than a third of the way down the list of matters to be discussed....

Having negotiated the minefield of the arrangements process there is then the day of the funeral itself to contend with. I run a small firm now, so funerals have to be carefully scheduled around each other; but back when I was with my original employers the size of their company was such that running three or four funerals a day in close parallel was very much the norm, with all the organized chaos that volume of work could sometimes entail behind the scenes. With two or three fully-laden hearses in the yard all due out at roughly the same time, there was always the potential for problems and although I can record with absolute certainty and honesty that we never had any mix-ups with coffins, we did have a couple of close shaves when the respective funeral directors and their teams of bearers nearly got in the wrong hearses.

Nevertheless, the entire twenty five years of my career so far have still been punctuated by occasional "day of the funeral problems," such as cars that decided to manifest flat batteries at the start of the working day, for example. On a somewhat more silly note, I very recently had to beg a hearse and limousine from my competitors because a power cut had disabled my electrically operated garage doors. The emergency manual override on the doors also proved problematic, leaving my vehicle fleet, all clean and shiny, stranded in the garage. The power was later restored and the problem with the manual override was also quickly fixed, but not in time for that day's funeral.

The journey to the venue of the funeral is the next hurdle to be overcome. There is a balance to be struck, because whilst being late is a disaster, I also regard arriving too early as being a very unpolished approach too. To my mind the coffin represents, in one sense or another, the guest of honour and it really doesn't seem right that it should be kept waiting outside the church or crematorium longer than is necessary. There should be something rather transcendental about the coffin at a funeral and it should be treated as such. Leaving the coffin waiting in the hearse for more than a few minutes, all the while with people stood around chatting and waiting for the service to start, to my mind detracts from the sense of occasion and diminishes the ritual.

So with that thought in mind, we must plan to allow for anything that will disrupt our careful journey timings. The most obvious threat is of course roadworks. Large scale roadworks usually occur with some degree of warning, so allowance can then be made, to one extent or another. However, temporary traffic lights and "pop-up" roadworks are an altogether more unpredictable menace to which we must constantly be aware of.

Staying with the subject of incidental hold-ups, whilst refuse collection lorries can be a nuisance, kerbside re-cycling lorries represent mobile tyranny for us. At least if you're stuck behind a dustbin lorry, you know it will keep moving as the collectors run up the road behind it, chucking the bin bags in. Kerbside re-cycling collections however, are a long, painful, inefficient process in comparison and a nightmare if you get stuck behind one. We had one such incident on the way to Cheltenham Crematorium - the most awkward to travel to of the three crematoria that we use regularly. The problem was made worse because the road in Stroud where we got held up is notoriously bad for parked cars blocking up the traffic at the best of times and the recycling lorry was not only holding us up but causing gridlock fore and aft as well.

We eventually broke free of this slow moving menace, scraping a wing mirror on the hearse whilst trying to squeeze past, before having to travel at a rather quicker pace for the rest of the journey to ensure that we arrived at the crematorium on time. The lady conducting the funeral service was a lay preacher and we had given her a lift with us in the hearse. She later took great delight in telling the family what had happened on the way over, much to my embarassment, as I take the view that on such occasions what the eye doesn't see the heart doesn't bleed over. But as it turned out, the family thought it was hilarious. Apparently their late mother always used to get frustrated with their father because he drove too slowly and timidly. They were delighted that mother was transported on her final journey with "the kind of assertive driving she would've approved of!"

In the case of crematorium funerals in particular, having arriving safely and on time there will still be further potential for problems. I'm not sure what invisible force governs crowd dynamics at funerals, but even something as seemingly simple as marshalling your mourners into and out of the chapel can be fraught with difficulty. Very often there is a collective unwillingness to be the first to move – everyone waits for everyone else, after which there is then a reluctance, despite clear invitation from the funeral director, to sit anywhere near the front of the chapel. I have had funerals where the rear of the chapel is packed, but there is then a gulf of emptiness between all the other mourners and the family group sat at the front.

There is also another phenomenon, unique to crematorium funerals and I remember one funeral in particular where this happened:

All the mourners had been seated and the service began with the usual opening words from the minister:

"I welcome you all today to this service of thanksgiving for the life of Jack Smith."

From where I was standing at the back of the chapel, I noticed two brassy-looking women in the congregation suddenly exchange embarassed glances with each other. With their gold hoop earrings, faux fir shrugs and dyed hair I'd already thought they looked very out of place amongst the other more conservatively dressed mourners on this funeral.

I watched as the two women fidgeted uncomfortably for another five minutes until, during the singing of the first hymn, they stepped out of their pew. It was at that moment I noticed they were both wearing skirts designed more for wearing in a nightclub than a crematorium. The crematorium attendant – a man with a constant roving eye for the ladies - was stood there transfixed, his eyes standing out like a pair of chapel hat pegs.

The two women teetered down the aisle on their six inch heels, with their heads held down, clearly determined not to make eye contact with any of the other mourners. One of them adopted a half crouched, waddling walk as she tugged at the hem of her mini-skirt, trying to pull it down to a slightly more modest eight inches above the knee. As she made for the chapel door she glanced up at me very sheepishly and hissed "Sorry love. We got the wrong funeral!"

Having got to the crematorium safely and on time, successfully marshalled the crowd of mourners together and seated them in the chapel with the minimum of fuss we can then be left with a comfortable margin of time for the service itself to take place. So does that mean that the funeral director is now on the home straight? No, not quite. Not if a family member, friend, or colleague of the deceased is to give a personal tribute.

I have watched, listened and cringed as some people have stood up and found the whole experience of public speaking so daunting that they have withered like an autumn leaf, before stumbling and muttering unintelligibly through their words.

I have watched helplessly as others have simply been overcome with emotion and either valiantly forced their words out through a quivering bottom lip or simply dissolved into tears and been unable to carry on. There is no shame in that; indeed on every one of those occasions the wave of sympathy and admiration coming from the congregation was palpable.

I have also watched, listened and enjoyed those times when tribute-givers rise admirably to the occasion, standing and speaking with confidence, clarity, humour and sincerity, always to the delight and appreciation of all present.

But there are also those who stand and speak with confidence, clarity and a total disregard for the instruction to speak for no more than five minutes, all the while showing utter disregard for the boredom threshold of their audience too. They start with the deceased's birth, working their way through every moment of the person's life in excruciating detail, before realising that they're enjoying the sound of their own voice so much they decide to turn the whole tribute into a solo act. These people have the power to reduce the crematorium schedule to chaos, as well as making life extremely difficult for the minister or officiant who subsequently has to cut large chunks out the service just to prevent it overrunning and delaying the next service.

Now, let's assume that we've got through the whole funeral so far without any hitches. Surely the end of the service means an end to the challenges to be faced by the funeral director when conducting a funeral? No, not really. Leading everyone out of the chapel is also a fine balancing act. The challenge is to ensure the crematorium chapel is vacated as promptly as possible, out of consideration to the mourners on the next funeral. Inevitably though, the family on our imaginary funeral may wish to linger near the exit door to meet and greet everyone. This can lead to a bottleneck in the chapel and many are the times my own funerals have been held up because the funeral in front of us has overrun, due to the sheer number of mourners creating a delay in clearing the chapel. Handling bottlenecks of departing people on a funeral requires a mixture of tact and firmness, because whilst large groups of mourners often show as much willingness to move as ketchup in a bottle, we must still try hard not to make them feel like they're being herded like sheep.

Assuming the entire funeral has passed without incident and all the many hurdles have either been negotiated successfully or, better still, have not occurred at all, is there finally some peace for the funeral director when he or she gets back to the office? What do you think...?!

We had one family who, for the sake of a grand-daughter due to fly away on holiday, asked if we could arrange the funeral for the day before her flight was booked. Not only did this request leave us with an uncomfortably short time frame in which to make the arrangements, but the funeral also had to be shoe-horned into a day when there were already two funerals timed in such a way that performing a third funeral was going to be difficult. Nonetheless, with a great deal of planning, shuffling, goodwill from doctors and registrars to obtain the necessary certificates and a total reliance on nothing unforeseen happening on the day, we managed to accede to the family's request.

But barely half an hour after we'd returned to the office following the funeral, the phone rang with a very irate young woman at the end of the line. It was the grand-daughter, furious that the flower card with her personal message was missing from the family floral tribute. A quick investigation with the florist revealed that the florist's staff had forgotten to put the card in the flowers. It was an awful mistake, but an honest mistake nonetheless and the florist immediately phoned and apologised to the grand-daughter, who promptly gave the florist a blast of verbal abuse for her trouble. At times like that one simply has to grit one's teeth...

It reminds me of a time when I was running my first business, Lansdown Funeral Service in Stroud, when I had to chase up another errant grand-daughter for settlement of the invoice for flowers we had ordered on her behalf for her grandmother's funeral.

The long overdue amount was just £20.00 and I decided that as reminder letters hadn't worked and the amount hardly warranted formal recovery action, I would instead discreetly let the grand-daughter's mother (who arranged the funeral) know that her adult daughter still owed us for her flowers. The grand-daughter's cheque arrived just a couple of days later accompanied by a very snotty letter giving not the slightest apology, but instead having a go at us for telling her mother!

I can only imagine what her mother must have said to her, but after leaving a £20.00 unpaid bill for flowers for her own grandmother's funeral I expect it was all the things I would quite like to have said...

CHAPTER NINETEEN

The Old Order Changeth

"The years teach much which the days never knew."
Ralph Waldo Emerson

It was Boxing Day 2009, during the Christmas that never quite happened. Well, not for me anyway.

Gloucestershire was experiencing its first really hard winter for over twenty years, with snow and ice the like of which we had not seen since 1982. I'd just arrived at my parents' house for lunch when my mobile phone rang. It was Polly, a member of my staff, who had already tried phoning half an hour earlier. The first time she called I'd had my hands full just trying to de-ice my frozen car and so with a windscreen scraper in one hand I had looked at my phone's display screen, saw it was Polly, assumed she probably just wanted to say "Happy Christmas" and rejected the call, intending to call her back later.

When Polly phoned again I was just reversing my car onto my parents' driveway, so I left Polly on the line while I parked. When I turned my attention back to the conversation Polly sounded very upset.

"It's my Brother-In-Law, Steve. You've met Steve, remember? At that birthday meal back in the Summer?" There was a pause and I could hear Polly struggling to compose herself. "He's committed suicide." She blurted out. "He's been found hanging in a park near his home. Heidi (*Polly's sister*) is desperate to see him, we all are. Can you arrange something, anything? Please?!"

I mentally processed the information as quickly as she gabbled her words out. Polly was in such a state on the phone that I simply did what came naturally to me and switched straight into funeral director mode. I tried to slow her down a bit and took her back to the beginning again, gently but firmly asking questions to establish just precisely what Polly knew at that stage.

I did remember her brother-in-law quite clearly, although I had only met him that one time, at a birthday celebration for another member of Polly's family.

I was able to ascertain from Polly that her brother-in-law had last been seen by the family on Christmas night, so I guessed he must have died in the early hours of Boxing Day morning. An unfortunate dog-walker had discovered his body. I knew the town where Steve lived and I was familiar with the area of parkland bordering the town where apparently his body had been found. The parkland was part of a country house estate and often yielded the bodies of local people who'd gone missing, usually with the intention of ending their own lives. My first experience of such a case was in the earlier years of my career, when my former employers had arranged the funeral of a young gamekeeper on the estate who had turned his gun on himself.

I cautiously promised Polly that I would do all I could to try and arrange for her sister Heidi to see her husband's body at the public mortuary later that day. By this time I was anxious to finish the call so that I could firstly gather *my own* thoughts, before doing anything else. I sat there in my car for five minutes working out how I would deal with their request, not to mention how I would slot it all into the now tattered remains of my parents' plans for Boxing Day. I went in the house, got all the parental greetings out of the way and quickly retreated outside into the freezing back garden where it was at least quiet enough for me to start all the phone calls that I was suddenly required to make.

Back then there were two public mortuaries serving the county, located at Gloucestershire Royal and Cheltenham General Hospitals respectively, with each mortuary's catchment area broadly aligned with that of its respective hospital. To begin with, the town where Steve had died was served by the Cheltenham mortuary. I phoned the hospital and asked them to get the duty mortuary technician to phone me back. When the return call came I was relieved and pleased to find it was Karen on duty at Cheltenham that day. I got on very well with Karen and I knew I could count on her co-operation. By this stage in time, both the county mortuaries had an all-female staff of technicians – Anatomical Pathology Technicians to use the proper term. Instead I jokingly referred to them as the PMT's (Post Mortem Technicians… of course!).

Fate had taken a hand as Karen was already at the mortuary, preparing for a forensic post-mortem later that afternoon, arising from a death that had occurred on Christmas Day, so she was able to check the register and inform me that Steve's body had not even arrived at the mortuary. It later transpired that Polly had been so quick in phoning me that the local undertakers had not even completed the removal.

I knew the mortuary had specific out-of-hours viewing times and there was no way that Steve's body would be received in time to fit in with those hours. However Karen, bless her, suggested a solution which I too had thought of: that if she was going to be there anyway then I could conduct the viewing whilst she concentrated on dealing with the forensic p.m. We agreed on a time of 6.00pm that evening, which allowed me to salvage some of my Boxing Day and would give Karen time to go home herself, before she headed back into work to meet the Home Office pathologist and the police for the forensic p.m. that afternoon.

Acting on karen's instructions, I asked Polly and her family to meet me in the reception of Accident & Emergency, as all other visitors to the mortuary were requested to do. A & E is only a very short walk away from the mortuary.

I phoned Roger Wade, the coroner's out-of-hours officer to ensure that he had no objection to our plans, knowing full well that strictly speaking formal identification should always take place before any routine viewing. However, Roger too was also very co-operative and we agreed that the most sensitive approach would be to allow Steve's wife Heidi to see his body, whilst Roger would meet us at the mortuary shortly after, for Heidi to then confirm the formal identification while she was there. Roger Wade, one of life's great pragmatists, was actually quite pleased with this arrangement, as it meant one less on the list of identifications to arrange after the Christmas period and consequently one case at least that could be processed quickly.

To drive to Cheltenham General Hospital's mortuary you had to negotiate a veritable maze of internal access roads at the rear of the hospital, which was ironic when you consider that the mortuary was actually located right on the very edge of the hospital campus, at one end of the Pathology Block. Undertakers used the mortuary's main entrance of course, discreetly hidden amongst other hospital buildings. Yet the entrance that relatives used, allowing access only to the viewing suite, was reached by an open path on the outer side of the pathology block, overlooked by a terrace of Victorian villas in the leafy residential street bordering that side of the hospital campus. From the street the Pathology Block looked so anonymous and unremarkable that I often wondered whether the residents of that terrace realised just what the hospital building opposite them actually housed.

My company's various suppliers (coffin manufacturers, etc.) always gave us bottles of wine at Christmas, along with the obligatory calendar and I usually passed the bottles on to other people. Giving a mainly non-drinker like me a bottle of red wine is a bit like fitting an ashtray to a motorbike, but I thought taking one of those spare bottles to the mortuary was the least I could do by way of a thank you to Karen for her co-operation. I arrived at the mortuary in good time but there was little I could usefully do by way of preparing Steve's body for viewing because, with the post-mortem still to be held, the pathologist would need to examine his body in the exact condition in which it was found and I couldn't risk disturbing any useful evidence (including, unfortunately, the ligature still in place around his neck).

Fortunately he looked very peaceful anyway, so with a hospital blanket pulled right up under his chin Karen and I wheeled him through into the viewing room – a tatty, bland space with outdated beige furnishings and ornamented like a chapel of rest. We still had half an hour before I was due to go and meet the family, so we retreated back into the main part of the mortuary. The forensic post-mortem had actually been completed by then and Karen was waiting for the Home Office pathologist and the police detectives to finish their de-brief in the PM room and leave her alone to get on with cleaning up instruments and washing everything down.

With that spare time to fill, Karen asked me if I could do anything to free up a jammed lock mechanism on one of the fridge doors, that was currently imprisoning the three bodies stored behind it. So, armed with a screwdriver that Karen had produced from somewhere, we chatted idly whilst I dismantled the lock.

Karen was keeping one eye on the pathologist and the detectives through the open PM room door. I could tell from the glances she was giving them that she'd decided they had outstayed their welcome. It had obviously been a long day for her.

My voice muffled from having my upper body squeezed into the opening of another fridge doorway adjacent to the jammed door, I remarked that it was "Flaming typical that you go and get lumbered with a forensic, today of all days."

"Yeah," replied Karen, "the kids are all excited with their presents so I expect my husband will be clawing at the walls by now!"

I laughed from within the fridge, where I was trying to hold the lock mechanism from the inside with one hand whilst twisting a screwdriver into the outer side. At that moment the pathologist and the C.I.D officers came back into the body store, ready to leave the mortuary. They were unaware that I was even in the mortuary and looked somewhat taken aback when the top half of me suddenly reappeared out of the open fridge, screwdriver in hand. I just smiled at them and carried on.

The pathologist and the detectives thanked Karen and cheerfully wished her a happy remainder of Christmas as she saw them off the premises. Holding the dismantled lock in my hand I watched as Karen said goodbye to her visitors. She shut and locked the front door behind them in a very firm manner that implied that even if one of them had accidentally left his car keys behind they probably wouldn't stand a cat in hell's chance of being let back in to retrieve it until the mortuary officially re-opened after the holiday period. Karen marched back into the body store with a look of grim satisfaction and we continued our conversation.

"So you won't actually get any more peace when you *do* get home then?!" I remarked.

"Not flaming likely." Came the weary reply.

"Bit rotten isn't it though. Being stuck in this place, up to your elbows in some poor unfortunate on the p.m. table, when you could be at home in the loving bosom of your family." I said jokingly.

"I'd only be chasing round after each of them, feeding them, keeping them all happy and refereeing arguments between the kids! Anyway, when I get back my husband will have to go out and start his shift at work. Once he's out the door he won't have it any easier than I have today."

Knowing Karen's husband ran a taxi business, I said, "No, I imagine he won't. Too many drunken idiots around at night as it is. I bet it's a nightmare for him over the Christmas and New Year period isn't it?"

"Ohhh yes." She replied, in a knowing tone of voice. "I've been out in front of our house in the early hours in my dressing gown and slippers before now, cleaning up vomit from the back of his taxi." she said, without the slightest hint of rancour.

I had always found Karen to be a very friendly and inoffensive soul but those words gave me a hint of a "don't mess with me" side to her nature that I'd never before suspected. I thought back to the way the Home Office Pathologist and the CID officers had just been politely but firmly seen off the premises. I decided a change of subject was in order.

"So, is this forensic likely to turn into anything interesting?" I enquired, knowing better than to ask, but asking anyway.

"I think they're treating it as an unexplained death rather than outright suspicious at the moment. Someone fell from a building. That's all I can say."

"Ah. A fudgy-wop case then." I remarked, ducking my head back in the fridge to do battle with the lock mechanism again.

"Do what?" I heard Karen ask, sounding puzzled.

"Fudgey-wop." I shouted from inside the fridge. "The great dilemma that keeps coroners in a job. F.J.O.W.P. - fell, jumped or was pushed!"

"Oh right, yeah!" Karen laughed.

It was time to walk across to meet Polly's family. Just as I reached the forecourt outside A & E, I spotted Polly, Heidi and their mother Barbara through the glass doors of an entrance adjacent to the emergency reception area. I pulled the door open and stood in the doorway for a second, looking at the forlorn little trio sat on plastic chairs in the drab, empty corridor. As the door swung closed behind me Heidi instinctively got up and walked towards me, a look of pleading bewilderment across her tear-stained face, as if she was hoping that whatever I had to say might shed some light on why her world and that of her two young children, had just disintegrated around them. In fact I didn't really know what I was going to say.

"Why did he do it?" She whispered, in part asking me and in part directing the question into the air.

"I don't know Heidi. I really don't know." I replied. "Is it the three of you, is anyone else with you?"

Heidi's mother, Barbara, spoke up. "No, it's just us." Then with a firm, resolute tone in her voice she asked, "Where do we have to go?"

"It's not far. I'll take you across."

I led them outside and across towards the Pathology Block and round the outer path to the mortuary viewing suite. I was effectively taking the place of a coroner's officer at that moment and despite all the hundreds of funeral home viewings I had done over the years, I had never done *this*. I was relying on experience and intuition to be my guide through that dreadful procedure. As we approached the bright red door of the viewing suite I suddenly noticed how exposed to public view we all were. Heidi, Polly and Barbara were all well beyond caring about their surroundings, but being out on the edge of the hospital campus, away from the protection of other buildings and in full view of what was a normal, residential street, I felt that this most private of moments was somehow far more public than it should ever have been.

With the four of us gathered in the lounge area – as bland and uninspiring as the rest of the tired viewing suite, I paused for a second before explaining as gently as I possibly could that I wasn't allowed to interfere with any evidence on Steve's body, which was why the blanket would be pulled up to his chin. I could tell that Heidi immediately understood my subtle hint and she reacted with both remarkable understanding and courage. She indicated a wish to go in alone to begin with so, with nothing more needing to be said, Polly and Barbara waited whilst I took Heidi through into the viewing room.

Later, as I stood by the viewing room doorway talking quietly to Polly, whilst Heidi and her mother were huddled over Steve's body, I heard a familiar voice coming from the lounge area. Leaving the three of them alone in the viewing room, I slipped out of the room to find the familiar figure of Roger Wade, every inch the ex-copper, stood waiting in the lounge, briefcase in hand. I brought the grieving trio out of the viewing room so Roger could speak to them all, before he would take Heidi back into the viewing room to confirm the formal identification. With the introductions done, I told Heidi, Polly & Barbara that I had to defer to Roger's authority as the coroner's officer and that for reasons of confidentiality I couldn't be party to their conversation with Roger. I told them I would go and wait outside.

With the identification formalities completed Roger emerged from the building, telling me that Heidi and Co. wanted a few more minutes with Steve's body. As we stood outside in the freezing night air he asked, in his usual business-like manner,

"D'you know if it will be burial or cremation?"

I smiled inwardly, remembering one coroner's removal years before when I had watched Roger bluster his way through an explanation of coroner's procedure to a woman who had found her elderly father dead in his bed barely an hour before. The poor woman could barely register the fact her father had died so suddenly after they'd spent an otherwise normal and happy day together and she was in no fit state to listen or even care about coroner's procedures at that moment. I remembered apologizing to the woman on Roger's behalf when he'd left us alone, explaining to her that his intentions were good, but that he was an ex-policeman and he sometimes got so wrapped up in procedure that he forgot about people.

Turning my mind back to the current situation, I explained to Roger that I didn't anticipate Heidi being in any fit state to even start thinking about funeral arrangements for a while and promised him that I'd phone the coroner's office the moment I knew anything myself.

"It's just that we can get the paperwork done a little quicker if we know" he pressed.

I knew he meant well, that he was anxious to speed matters up for the family's benefit.

"I appreciate the thought Roger," I replied tactfully, "But to be honest I really don't think they'll be in too much of a hurry. They've got enough to think about as it is."

"Ok, well keep us informed and we'll get straight onto it as soon as we hear from you." With that he bade me a cheerful goodbye and strode off into the night.

I stepped back into the lounge of the viewing suite just as Heidi, Polly and Barbara were emerging from the viewing room.

"Thank you ever so much for arranging this for us James. We know it's been a massive hassle for you. It's ruined your Boxing Day." Polly said.

"That doesn't matter Pol', this is the least I can do, considering the circumstances." Then, gently directing my words to Heidi,

"Just remember there'll be plenty more opportunity to see Steve again in due course.... if you or the children want to."

I later learnt that Heidi and her parents had broken the news to her two young children earlier in the day. Now, with the two kids safely occupied with their Grandad and their Auntie Louise – Heidi's youngest sister, they were unaware that their mother was at the mortuary identifying their father's body.

"We'll be in touch soon, James." Barbara said, her face still fixed with the same firm, resolute expression she had worn from the moment we had met over in A & E.

Taking charge of her daughter once again, Barbara put her arm protectively around Heidi's shoulders and led her to the door. We said our goodbyes and after locking the viewing suite door behind them I made my way back through into the mortuary.

I helped Karen place Steve's body back in the fridge and as we walked through into the entrance vestibule she unlocked the front door to let me out. Standing out in the vehicle bay I reached into my pocket for my car keys.

"It's been a very brief Christmas this year, hasn't it?" I remarked ruefully.

"Certainly has." Karen sighed. "I'm on call again tomorrow as well, so I'll have to be back in for the formal ID on the forensic case, then I've got a feeling there'll be a lot of viewing requests on that one. I'll be bloody glad when the coroner releases that body. The funeral directors can deal with it all after that"

Knowing what I too was likely to have waiting for me in the coming days, I said "We've both got a trying time coming up then."

Karen gave me a wry smile and closed the mortuary door.

Formal identifications are no longer carried out at Cheltenham General Hospital's mortuary. Along with its counterpart mortuary at Gloucestershire Royal Hospital, they've been replaced by a £4.9m purpose-built Coroner's Court and County Mortuary in Gloucester. The complex, which can accommodate sixty bodies, deals with the twelve hundred deaths reported to HM Coroner for Gloucestershire each year. Karen and her technician colleagues have transferred there and the two hospital mortuaries are now downgraded simply to body storage and viewing facilities, handling only routine hospital deaths.

Meanwhile all the old "Rose Cottage" mortuaries at the smaller hospitals, of which there were five that I used to regularly go to, have all gone completely; either because the hospitals themselves closed or because we now have to endure the iniquity of removing bodies directly off the hospital wards. This procedure either means totally unnecessary night call-outs for already tired and stressed funeral staff, or during the day the utterly repugnant business of wheeling bodies around in full view of patients and visitors to the hospitals. The Hospital Trusts in question justified closing their mortuary facilties by saying they were too expensive to refurbish to an acceptable, modern standard or, in one case, that following a multi-million pound refurbishment no space could be found for mortuary facilities to be housed. Neither excuse deserves the dignity of any further discussion here.

Thomas Broad & Son still exists, but only in name now. After its takeover by a large funeral group none of the original staff remain, save for a few of the part-timers. The funeral home itself - Dartford House, still stands of course, in the seclusion of its grounds, although internally it no longer bears any relation to the funeral home that I knew so well. Everything has been remodeled; the office and chapel of rest both altered and rebuilt, whilst my old haunt, the coffin workshop, was also sacrificed in the refurbishments and relocated to an outbuilding elsewhere in the yard. The mortuary is, I'm told, the only part of the funeral home unchanged from when I was there, but Stan, one of the original remaining part-timers, told me that it's not kept as clean and tidy as I used to keep it.

My former bosses, Paul & Michael, are both retired now, whilst my former colleague Rick now holds a senior position in Gloucester's largest privately-owned funeral company. He is also now an accredited Humanist Funeral Officiant. T'other James remains with me at Fred Stevens Funeral Directors and we are now two of the longest serving funeral directors in the Stroud district. Other young undertakers, who started out as teenagers at about the same time as me and who I've got to know over time, are themselves now either in senior positions or have taken the helm of their respective family businesses.

It might sound as though everything that I have described in this book has gone, has moved on or has ended and to some extent that's true, but that's why I have written about all those things now. The profession I joined was just on the cusp of change and I'm glad I was able to see how things used to be.

Funerals themselves are changing. The comfort that was once gained from conformity to established traditions and communal rituals is being challenged by an ever more individualistic society. Personalisation is everything now and public attitudes have changed too, with people seeming ever more keen to embrace the rituals of death and rewrite them as their own. This is a time of flux for funerals and it's still too early to fully discern the shape of things to come. There is no doubt however, that the funeral profession will, like funeral rituals themselves, be reinvented to one extent or another in the coming years and it will be interesting to see how much things change and how much they stay the same. However, whilst the requirements made of funeral directors may change, I think people will always want the assistance of capable, experienced professionals nonetheless.

The one thing I can be sure of is that I've been extremely fortunate to have had such a thorough training and grounding in funeral directing. The next quarter century will, no doubt, be every bit as eventful as the previous one and I'm sure the challenges will be very different, but I always think the true value of a journey lies in knowing where you've come from, more so than in knowing where you are going to, so in that sense I feel I'm well prepared.

"And in today already walks tomorrow."

Samuel Taylor Coleridge

Lightning Source UK Ltd.
Milton Keynes UK
UKHW040657260419
341665UK00001B/328/P

9 780957 346802